The History of the Small Block Chevy Motor

Todd Bandel

Copyright © 2024 Todd Bandel

All rights reserved.

ISBN: 9798344246536

DEDICATION

I dedicate this book to all my past automotive mentors and colleagues. Your guidance, support, and shared wisdom have been invaluable in shaping my journey. Each of you played a significant role in my professional development, imparting knowledge and fostering a passion for excellence in the automotive field.

CONTENTS

ACKNOWLEDGEMENT   i

Chapter 1
Conceiving the Small Block Chevy   *1*

Chapter 2
Anatomy of a Legend   *13*

Chapter 3
Generations of Power – Evolution and Transformation   *31*

Chapter 4
Revolution on the Assembly Line   *45*

Chapter 5
The Heartbeat of America's Passenger Cars   *67*

Chapter 6
Track Domination, A Racing Dynasty   *91*

Chapter 7
The Canvas for Custom Culture   *107*

Chapter 8
The Small Block Goes Global   *123*

Chapter 9
The Pop Culture Engine   *143*

Chapter 10
The Enduring Legacy and Future   *157*

ACKNOWLEDGMENTS

I want to express my deepest gratitude to my father for introducing me to the exhilarating world of automotive racing. Your passion for cars and dedication to the sport have inspired me.

From the first time you took me to a race track, I was captivated by the power and precision of the machines, as well as the skill required to master them.

Your guidance and support have fueled my interest and enthusiasm, making every moment in this thrilling world more meaningful. Thank you for sharing this incredible journey with me and for being such a pivotal influence in my life.

Chapter 1: Conceiving the Small Block Chevy

Picture a smoky engineering lab in the early 1950s. On one side of the room, an aging inline-six wheezes its way through another dyno pull. On the other hand, a small group huddles over fresh blueprints for something radical: a compact, lightweight V8 that might not just save Chevrolet's reputation, but change the trajectory of American car culture. They have fifteen weeks to prove it can work.

This opening chapter sets the stage for everything that follows. Before we can appreciate the Small Block's racing glory, its swap-meet ubiquity, or its cultural mythology, we need to understand the conditions that forced Chevrolet to rethink what an engine could be. Here, we anchor the story in post–World War II America, introduce the key figures, especially Ed Cole, and trace how a tight timeline and real-world manufacturing constraints shaped the design in ways that made it endlessly adaptable later on.

Section 1.2: Post-War America at a Crossroads

The Small Block Chevy did not come into existence in isolation. It was a product of a distinct period in American history, shaped by increasing prosperity, expanding highways, and shifting consumer demands that rendered the traditional inline-six less sufficient.

In the early 1950s, American buyers wanted more than basic transportation; they wanted capability. Families were driving farther, faster, on new interstate highways that stretched across the country like steel ribbons. They expected smoother acceleration, higher cruising speeds, and enough power to haul growing cars without feeling strained. At the same time, competitors were raising the bar. Oldsmobile and Cadillac within GM, and other makes outside it, were already fielding modern overhead-valve V8s that made Chevrolet's offerings look positively quaint by comparison.

Chevrolet's workhorse inline-six, reliable as it was, simply couldn't keep pace with the performance and refinement expectations of this new era. It was long, tall, and limited in its breathing and ultimate output. Executives began to understand that if Chevrolet didn't evolve, it would quickly look outdated in its own showrooms, a particularly painful prospect for a division that prided itself on volume leadership.

Yet any new engine had to do more than just make power. It had to be affordable, durable, and buildable at enormous volume. GM was not chasing an exotic halo motor; it needed a new backbone for the brand, something that could power everything from family sedans to light trucks without breaking either the budget or the assembly line.

Consider a typical 1950 Chevrolet sedan: solid, sensible, but modest in acceleration and top speed. Now place it next to a contemporary Oldsmobile Rocket V8–powered car. The difference in responsiveness and confidence at highway speeds was obvious to drivers and magazine testers alike. That comparison played out in showrooms all over America, and dealers felt it in their order books.

Period road tests and dealer feedback regularly noted that buyers were asking for "more power" and "easier passing" on open roads. The message to Chevrolet's engineering staff was clear: the company could not rely on incremental tweaks to a pre-war inline-six architecture indefinitely. The market was moving, and Chevrolet needed to move with it, or risk being left behind.

When you look at major shifts in automotive history, how often do you see them driven more by changing driver expectations than by purely technical curiosity? The Small Block's story is fundamentally one of market forces creating opportunity. Think about a modern vehicle segment you follow. Where do you see a similar tension today between legacy designs and emerging consumer demands? The parallels are often striking.

If you have access to period road tests or online archives, spend ten to fifteen minutes reading 1950–1954 reviews of Chevrolet versus its V8-equipped competitors. Make a short list of the complaints and praises related to engines. Note how often words like "power," "smoothness," and "economy" appear together; those competing demands will echo throughout this book, because they shaped every decision Chevrolet's engineers made.

Section 1.2: Ed Cole and the Visionary Engineering Team

Behind the Small Block's creation stood not just a company, but a core group of engineers, led by Ed Cole, who combined practical manufacturing sense with an almost stubborn belief that a compact, efficient V8 could serve both the average driver and the performance enthusiast.

Ed Cole came to the project with valuable experience from Cadillac and prior V8 development. He understood not only how to design an engine but also how to get it approved, built, and sold at scale within a large organization. His leadership style emphasized clarity of purpose: the engine had to be light, compact, inexpensive to build, and flexible enough to serve a wide range of vehicles. Cole had the kind of vision that sees not just the immediate problem, but the downstream implications, how a design choice made in 1952 might enable or constrain possibilities in 1962.

The History of the Small Block Chevy Motor

Supporting Cole were engineers like Zora Arkus-Duntov and John Dolza, each bringing their own strengths to the table. Arkus-Duntov, with his European racing background, would later become synonymous with Corvette performance and helped push Chevrolet to recognize the Small Block's high-performance potential from the start.

He understood that an engine designed purely for economy and reliability would miss opportunities, that with the right architecture, you could have both mainstream appeal and enthusiast credibility. Dolza contributed in key areas of design and analysis, helping to translate ambitious goals into workable geometry and component choices, turning conceptual sketches into parts that could actually be cast, machined, and assembled.

What unified them was a willingness to challenge established assumptions. Instead of simply scaling up existing ideas or copying what competitors had done, they looked for ways to reduce weight, simplify casting, and improve breathing, all while keeping costs under control. They asked fundamental questions: Why does the block need to be this thick? Can we route oil passages more efficiently? What if we reconsider the entire valvetrain layout?

Internal GM correspondence and later interviews with Cole's team often highlight his insistence on a compact "package size." He drove home that the new V8 had to fit into existing engine bays and allow room for accessories and serviceability, which would directly influence bore spacing, deck height, and overall block dimensions. This wasn't abstract engineering; it was practical problem-solving with real constraints and real consequences.

Arkus-Duntov's famous "Thoughts About Youth, Hot Rodders and Chevrolet" memo, though written slightly later, captures his mindset perfectly. He saw the Small Block not just as a transportation unit, but as a platform for performance and image-building. That perspective pushed the team to think beyond minimum requirements and consider how enthusiasts and racers might modify, enhance, and adapt the engine, pushing it far beyond stock specifications.

The History of the Small Block Chevy Motor

When you think about influential engines you admire, how often do you also think about the people behind them? Does knowing their intent change how you view the hardware? In your own work or projects, where have you seen strong leadership make the difference between a "good enough" solution and a truly innovative one? The Small Block's story suggests that great engineering comes not just from talent, but from a shared vision that aligns individual contributions toward a common goal.

Take a moment to jot down three names associated with the Small Block Chevy: Ed Cole, Zora Arkus-Duntov, and John Dolza. Next to each, write a one-sentence description of what you understand their role to be. As you continue through the book, revise and refine these notes. You'll build your own quick-reference view of the human side of this engineering story, and you'll see how individual personalities and perspectives shaped the final product.

Section 1.3: Revolutionary Design Principles

The original Small Block's greatness didn't come from a single radical trick or one brilliant innovation. It came from a set of interlocking design principles: overhead valves, thin-wall casting, compact configuration, and efficient combustion chambers, executed with unusual discipline and attention to manufacturability.

The overhead valve architecture, while not entirely new to the industry, was optimized in the Small Block for a low, compact profile. Valves in the head allowed better breathing than flathead designs and set the stage for high specific output without excessive complexity. The pushrod arrangement was simple, serviceable, and reliable, critical factors for an engine that would see duty in everything from delivery trucks to sports cars.

Thin-wall casting represented a significant technical challenge. By pushing foundry techniques to reduce unnecessary mass in the block and heads, Chevrolet engineers created an engine that was lighter than its competitors without sacrificing strength. This required careful control of core design and cooling paths; too thin, and you risk hot spots or structural failure; too thick, and you give away the weight advantage that makes the whole exercise worthwhile. The reward was a lighter engine that still met durability targets, contributing to better vehicle handling and fuel economy.

The compact configuration came from tight bore spacing and efficient overall dimensions. This allowed the Small Block to fit easily into a variety of chassis and left room for suspension components, steering linkages, exhaust routing, and future accessory drives. In an era when engine bays were already crowded with power steering pumps and air-conditioning compressors, this packaging efficiency was more than a convenience; it was a strategic advantage.

The early wedge-style combustion chambers and carefully chosen valve angles helped create good swirl and mixture motion, supporting both power and fuel economy. The design struck a balance between simplicity to machine and effectiveness in real-world conditions. More exotic chamber designs might have offered marginal improvements in efficiency or power, but at the cost of machining complexity and higher rejection rates in production.

Each of these principles was evaluated not just for performance potential, but for how it would behave in casting plants, machining lines, and dealer service bays. That dual focus, engineering excellence plus production practicality, is one of the Small Block's defining traits. The team understood that an engine design is only as good as your ability to build it reliably, service it affordably, and adapt it to changing needs.

Compared to some contemporary V8s, the Small Block's block casting looked almost minimalist. Less material around the cylinders, shorter overall height, and streamlined external surfaces all contributed to a favorable power-to-weight ratio that enthusiasts would later exploit on track and street. The engine didn't look overbuilt, it looked purposeful, like every ounce of metal was there for a reason.

Dyno comparisons of early prototypes showed that even with conservative cam timing and compression ratios suitable for mass-market fuel quality, the engine produced strong torque and competitive horsepower. The architecture clearly had "headroom" for more aggressive tuning later on, a characteristic that would prove invaluable as performance demands increased and aftermarket support developed.

When you examine a successful mechanical design, how often do you look for the underlying principles instead of focusing solely on individual parts? Which of the Small Block's core principles, lightweight casting, compact size, and efficient combustion, do you consider most important? The answer likely depends on your perspective: a drag racer might prioritize one aspect, while a street rodder or restoration specialist might value another.

If you've worked on or rebuilt engines, where have you seen the tradeoff between ease of manufacture and ultimate performance play out in a tangible way? Choose one principle, say, thin-wall casting, and do a brief comparison: look up the approximate weight and external dimensions of an early Small Block Chevy and one competing V8 from the same era. Note the differences and imagine how those numbers would affect vehicle handling, packaging, and fuel consumption. The differences can be startling.

Section 1.4: Fifteen Weeks to Prove the Concept

The Small Block's initial prototype phase was conducted under intense time pressure, roughly 15 weeks from clean-sheet concept to running engine. That constraint forced the team to focus on what truly mattered, driving rapid iteration and ruthless prioritization of features that delivered real benefit.

With corporate expectations rising and competitive pressure building, Chevrolet's engineering staff didn't have the luxury of endless experimentation. They needed a running, testable engine quickly, one that could demonstrate its potential to skeptical executives who controlled the funding for full-scale production. This accelerated schedule shaped the way the design was finalized: wherever possible, they leveraged proven ideas, simplified components, and minimized unnecessary complexity.

Technical challenges ranged from controlling core shift in thin-wall castings to ensuring adequate lubrication in a new block layout. Manufacturing engineers and design engineers worked in closer collaboration than was always typical, because any feature that looked elegant on paper but failed in the foundry or on the machining line could jeopardize the schedule. This forced integration of design and manufacturing, working side by side rather than sequentially, became one of the program's unexpected strengths.

The compressed timeline also meant that early test failures had to be treated as data, not disasters. Each problem, whether bearing distress, casting porosity, or cooling imbalance, was an opportunity to refine the design quickly and decisively. The team developed a rhythm: identify the issue, propose solutions, test, evaluate, and iterate. No time for perfectionism or second-guessing, just disciplined problem-solving under pressure.

Accounts from the program describe multiple prototype engines being built and torn down in rapid succession. Engineers would log issues, adjust dimensions or oiling paths, and have revised parts on the dyno in days, not months. That tempo is demanding even by modern standards, and it speaks to the team's focus and determination. They knew they had one shot to prove the concept, and they weren't going to waste it.

One recurring concern in thin-wall castings was maintaining uniform wall thickness to avoid hot spots and structural weak points. Solving this required careful pattern design and close communication with the foundry, an early example of design-for-manufacturability thinking that modern engineers now take for granted. The casting process had to be robust enough to produce consistent parts at high volume, not just one-off prototypes that worked under ideal conditions.

The History of the Small Block Chevy Motor

Section 1.5: Breakthroughs, Validation, and the Green Light

Prototype dyno sessions and durability tests produced the evidence Chevrolet leadership needed: this compact V8 could meet performance, reliability, and cost targets. Those breakthrough moments on the test stand translated directly into executive confidence and final approval for production.

Dyno testing verified that the new engine delivered strong torque across the usable rpm range, with horsepower figures that would be competitive in the mid-1950s market. Just as important, extended endurance runs began to demonstrate that the thin-wall block and compact bearings could survive real-world duty cycles. The engine wasn't just powerful on paper; it was durable in practice.

Engineers tracked oil pressure stability, bearing temperatures, combustion behavior, and coolant flow patterns during these tests. Where data revealed weaknesses, they iterated, adjusting oil galleries, revising water-jacket routing, or tweaking combustion parameters. Every test generated information, and every piece of information fed back into the design process. This empirical, data-driven approach gave the team confidence that they weren't just guessing; they were building knowledge.

Each successful test run didn't just add numbers to a chart; it reduced perceived risk in the minds of executives who would ultimately approve tool-up costs and production schedules. When performance, durability, and projected unit cost finally aligned, the program crossed the threshold from experiment to committed product. The decision to proceed represented a massive investment, retooling factories, training workers, and developing supply chains, and it required proof, not promises.

Reports from the era note that the new V8, even in its initial 265-cubic-inch form, showed a promising power-to-weight advantage over many competitors. For a mass-market Chevrolet, that was a compelling story: a lighter engine that could still move heavier, better-equipped cars with ease. The marketing potential was obvious, but it rested on a foundation of engineering reality.

Durability testing sometimes involved running engines at high load for extended periods, then tearing them down to inspect wear patterns. Seeing the main bearings and cylinder bores in good condition after such punishment was a key confidence booster for both engineers and management. It proved that the thin-wall block wasn't a fragile compromise; it was a genuine advance.

Section 1.6: Designing for Mass Production from Day One

The Small Block was engineered not as a boutique performance piece, but as a mass-production workhorse. From the outset, every major design decision had to reconcile performance ambitions with the realities of casting, machining, assembly, and long-term service. This approach made the engine both powerful and widely manufacturable.

Chevrolet needed an engine that could be built hundreds of thousands of times per year without crippling scrap rates or excessive machining time. This requirement influenced block geometry, core design, fastener selection, and even the layout of oil and coolant passages. Every feature had to pass a simple test: can we build this reliably, at scale, with the equipment and workforce we have?

Standardization and modularity were also key. The architecture had to support future increases in displacement and performance variants without requiring a complete reinvention of the casting and machining processes. That foresight would later enable the growth from 265 to larger displacements like 283, 327, 350, and beyond, each iteration building on the same fundamental platform.

On the assembly line, ease of access and straightforward assembly steps mattered enormously. Components had to be oriented so that workers and tools could reach them reliably, and tolerances had to be set within the capabilities of 1950s production equipment. An engine that was difficult to assemble would slow the line, increase labor costs, and create quality control problems. Simplicity wasn't just elegant, it was economical.

The decision to use relatively simple wedge combustion chambers and a conventional cam-in-block OHV layout helped control machining complexity. More exotic designs might have delivered marginal gains in efficiency or power, but at unacceptable cost and risk for a high-volume Chevrolet. The team chose proven solutions that could be executed consistently, day after day, shift after shift.

The basic bore spacing and deck height chosen for the original 265 proved flexible enough to accommodate later increases in displacement. That was no accident; it reflected a strategic decision to leave "room" for future evolution within the same fundamental architecture. The engineers understood that markets change, and an engine platform that couldn't adapt would have a short lifespan.

When you look at products or systems that have lasted for decades, how often do you see this same pattern, a core architecture designed from the start with future variants and mass production in mind? In your own projects, do you tend to design for a single immediate use case, or do you consciously leave space for evolution and scaling? The Small Block's longevity came from decisions made in its first fifteen weeks, decisions that valued adaptability as much as immediate performance.

Take a current project, mechanical, digital, or otherwise, and ask: "If I had to produce this 100,000 times, what would break first: the design, the process, or the cost?" This is the same kind of thinking Chevrolet's engineers applied to the Small Block, and it's a discipline that separates products that endure from those that fade.

In this chapter, we traced how post-war America's changing expectations exposed the limits of Chevrolet's inline-six and created the opening for a new kind of engine. We met Ed Cole and his core team of visionaries, examined the carefully chosen design principles behind the Small Block, and saw how a compressed fifteen-week development window forced focused, disciplined engineering. We followed the engine from fragile prototype to validated powerplant, and we highlighted the strategic manufacturing decisions that made it suitable for true mass production.

The guiding idea of this chapter is that the Small Block Chevy was born at the intersection of constraint and ambition. Market pressure, time limits, and manufacturing realities didn't stifle innovation; they shaped it, pushing Chevrolet's engineers toward an engine that was not only powerful and compact but endlessly adaptable. Understanding that origin story gives you a deeper appreciation for everything the engine would later become, from race winner to swap-meet staple.

Now that you've seen how the Small Block's foundational architecture came together, the next logical step is to open it up, literally. In the following chapter, we'll move from history to hardware. We'll dissect the engine's anatomy: block, heads, rotating assembly, lubrication, and cooling systems. You'll see how the design principles introduced here play out in cast iron and steel, and why the relationships between these components are so crucial for both stock reliability and high-performance builds.

For now, keep this image in mind: a once-experimental V8 on a 1950s dyno, barking through open headers while a handful of engineers watch the gauges climb and refuse to look away. They weren't just chasing horsepower; they were laying the groundwork for an engine that would power family sedans, sports cars, race winners, and backyard dreams for decades. In the next chapter, we'll see exactly what they built, one component at a time.

Chapter 2: Anatomy of a Legend

Imagine a bare Small Block Chevy sitting on an engine stand. No paint, no accessories, just cast iron and steel. At first glance, it's almost modest, compact, cleanly packaged, nothing obviously exotic. Yet builders who have spent a lifetime around engines will walk past far newer designs and stop at this one. They'll point to the deck height, the bore spacing, the way the lifter valley is laid out, and say, "That's just right."

This chapter explains why.

In Chapter 1, you saw how market pressure and bold engineering decisions brought the Small Block into existence. Now that the engine is "born," we turn from the story of its creation to the reality of its physical form. What exactly makes the Small Block Chevy's physical design so effective, and why did this particular combination of parts become the gold standard for adaptable V8 performance?

This chapter is pivotal because it establishes the mechanical foundation on which every later chapter builds. From generational evolution in Chapter 3 to racing dominance and custom culture in Chapters 6 and 7, understanding the anatomy now means the rest of the book will snap into focus. We're going to peel back the valve covers, so to speak, and look at how each major subsystem, block, rotating assembly, valvetrain, heads, lubrication, and cooling, works both on its own and as part of a tightly integrated whole.

By the end of this chapter, you'll understand the core architecture of the Small Block Chevy: block layout, bore spacing, deck height, and general dimensions. You'll see how the valvetrain, rotating assembly, and cylinder heads were engineered for both reliability and performance.

You'll recognize how lubrication and cooling choices contributed to durability and made high-RPM use practical. Most importantly, you'll be able to look at a Small Block, on a stand or in a car, and mentally "map" how its subsystems interact. You'll gain a practical sense of why certain modifications work beautifully with this design, while others run counter to its underlying logic.

Think of this journey as starting with the skeleton, then adding muscles, lungs, circulation, and finally temperature control. Each system matters individually, but it's how they work together that creates a legend.

Section 2.1: The Engine Block – Compact Strength by Design

The Small Block's engine block is the structural backbone of the entire design. Its thin-wall casting, compact external dimensions, and carefully chosen bore spacing created an unusually light, rigid, and adaptable foundation. This wasn't an accident or a matter of luck; it was deliberate, thoughtful engineering responding to very specific constraints.

Chevrolet's engineers deliberately targeted a compact V8 that could fit where an inline-six once lived. That meant strict limits on length, width, and weight, without sacrificing strength. Remember, post-war consumers wanted power, but manufacturers needed to control costs and maintain production efficiency. The solution lay not in adding material everywhere, but in placing it precisely where it mattered most.

They achieved this through several key innovations. First, thin-wall casting technology allowed them to use more precise casting techniques, reducing unnecessary material while maintaining rigidity in critical areas such as main bearing webs and cylinder walls. This wasn't about cutting corners; it was about sophisticated metallurgy and manufacturing control. The result was a block that weighed significantly less than competitors' designs without compromising structural integrity.

Second, the 4.40-inch bore spacing represented a critical design decision. This dimension balanced room for future displacement growth with a compact overall footprint. Too narrow, and you'd limit how much you could overbore cylinders or accommodate larger valves. Too wide, and the engine would grow too large for compact installations. The 4.40-inch spacing hit what engineers call the "sweet spot", a dimension that would prove remarkably prescient as displacement grew from 265 to 283, 327, 350, 400, and beyond.

Third, the deep skirt design extended the block below the crank centerline, bracing the main bearings and improving bottom-end strength. Picture the difference between a table with short legs and one with long, deep legs; the deeper structure resists twisting and flexing under load. In an engine, that translates to reduced vibration, better bearing life, and the ability to handle higher torque loads.

Finally, the integrated lifter valley created a centralized, rigid area that supported the camshaft and lifters while simplifying oil distribution. Rather than treating the lifter bores as separate features drilled into the block, the design incorporated them into the casting from the beginning, creating a structure that was both stronger and easier to manufacture consistently.

This architecture not only made the original 265-cubic-inch engine robust but also left room for future growth to 400 cubic inches and beyond in the aftermarket. Engine builders often remark that a production Small Block block can reliably handle power levels far above what Chevrolet originally intended, especially with minor reinforcement, an unplanned but welcome side effect of conservative engineering.

Compare a Small Block Chevy to a contemporary early-1950s V8 from another manufacturer. The Chevy block typically weighs less for a given displacement, occupies less space in the engine bay, making access and swaps easier, and accepts a wide range of crankshaft and bore combinations without requiring a redesign of the basic casting. These advantages weren't immediately obvious in 1955, but they would become increasingly valuable as hot rodding culture exploded and engine swaps became a defining characteristic of American car culture.

When you think about "a strong engine," do you imagine a thick, heavy castings, or carefully placed material where it actually matters? The Small Block challenges the assumption that strength requires mass. Instead, it demonstrates that intelligent design can deliver both strength and efficiency simultaneously.

How might Chevrolet's decision to prioritize compact packaging and thin-wall casting have influenced the hot rodding and engine-swap culture that followed? Consider this: a lighter, more compact engine is easier to handle, fits in more chassis, and leaves room for accessories and modifications. Those practical advantages created opportunities for enthusiasts that heavier, bulkier designs simply couldn't match.

Section 2.2: Valvetrain Architecture – Breathing Through Overhead Valves

The Small Block's overhead valve (OHV) valvetrain, pushrods, rocker arms, and a centrally located camshaft deliver efficient breathing in a compact, reliable package that responds exceptionally well to performance tuning. This design choice would prove central to the engine's adaptability and longevity.

The History of the Small Block Chevy Motor

Chevrolet chose a cam-in-block, pushrod OHV layout instead of an overhead cam design. At first glance, that might seem conservative; after all, overhead cam engines were already established in European sports cars and racing applications. But in practice, the pushrod design created a short, stiff valvetrain that was easy to manufacture and simple to service. Those practical considerations mattered enormously for mass production and long-term reliability.

The system works through a logical chain of components. A single camshaft mounted in the block drives hydraulic or mechanical lifters, which in turn drive pushrods, rocker arms, and finally the valves themselves. The straight, relatively short pushrods reduce flex and help maintain precise valve timing, especially at higher RPM. Unlike designs with long, complex linkages, the Small Block's pushrod path is nearly direct, meaning less energy is lost to deflection or compliance.

The stamped steel rocker arms used in most applications were inexpensive, durable, and easy to upgrade to higher-ratio or roller designs. This seemingly simple choice created an aftermarket ecosystem: enthusiasts could start with stock stamped rockers, then upgrade to roller-tip versions, then to full roller rockers with adjustable geometry, all while using the same basic mounting points and shaft configuration.

The symmetrical layout simplified oiling and manufacturing while maintaining good geometry across all cylinders. Unlike engines where cylinder banks receive different cam profiles or timing events, the Small Block treated all eight cylinders identically. This symmetry reduced manufacturing complexity and ensured that modifications were applied equally across the engine.

The result is a valvetrain that can deliver smooth idle, strong midrange torque, and, when paired with the right camshaft and heads, impressive high-RPM power, all within a small physical envelope. Consider how easily a Small Block responds to a camshaft change. Swapping from a mild factory cam to a performance grind can dramatically alter idle quality, torque curve, and peak horsepower, without touching the basic valvetrain layout. That responsiveness became a defining characteristic for generations of enthusiasts who learned that bolt-in modifications could transform engine character.

Racing applications have pushed pushrod Small Blocks to sustained high RPM for decades. Pro Stock drag racing engines routinely exceed 9,000 RPM. NASCAR engines run at 9,000-plus RPM for hours at a stretch. This long record of reliability under stress speaks to the inherent soundness of the valvetrain geometry. When properly executed, the pushrod OHV design delivers exactly what it needs to: precise valve control with minimal reciprocating mass.

Why do you think the pushrod OHV layout has survived in performance applications even as overhead cam designs have become common in modern engines? The answer lies partly in packaging: a pushrod engine is significantly shorter than an equivalent overhead-cam design, which matters for hood clearance and center of gravity. But it also lies in simplicity: fewer parts, fewer places for things to go wrong, and easier access for service and modification.

When you hear "simple" in an engineering context, do you tend to underestimate how much thought goes into making something appear that straightforward? True simplicity is hard-won. It requires understanding the problem so thoroughly that you can eliminate everything unnecessary and optimize what remains.

Look up a cutaway diagram or animation of a Small Block valvetrain in motion. Trace the path of motion from the cam lobe to the valve: cam lobe → lifter → pushrod → rocker arm → valve.

Note how few parts sit between the cam and the valve, and how direct the motion is. That directness represents countless hours of engineering refinement, determining the optimal angles, ratios, and geometry.

If you're planning a build, jot down how aggressive you realistically want your camshaft to be, daily drivability versus track use, and keep that in mind as we later discuss how the rest of the engine supports those choices. The valvetrain doesn't exist in isolation; it must work harmoniously with induction, exhaust, compression ratio, and intended RPM range.

Section 2.3: Crankshaft and Rotating Assembly – Smooth, Durable Power Delivery

The rotating assembly, crankshaft, connecting rods, pistons, and related hardware, translates combustion into usable torque. In the Small Block Chevy, this assembly balances strength, weight, and cost in a way that proved remarkably scalable from grocery-getters to race cars.

The original Small Block crankshaft design was conservative where it needed to be and efficient everywhere else. Five main bearings support the crank, providing stability and reducing flex under load. This was more than contemporary inline-six engines used, and the same count as most competing V8s, but the Small Block's compact dimensions meant those bearings were closer together, creating a stiffer, more rigid assembly.

The 90-degree V8 configuration with even firing intervals ensured smooth operation. Every 90 degrees of crankshaft rotation, a cylinder fires, creating a consistent, overlapping power delivery that minimizes vibration and maximizes smoothness. This wasn't unique to Chevrolet, but executing it well within such compact dimensions required careful attention to counterweight placement and balance.

The use of common journal sizes across different displacements later allowed easy interchange of cranks and rods across applications. A builder could start with a 350 block, install a 400 crankshaft, and create a 383 stroker combination. This interchangeability wasn't necessarily planned from the beginning, but it emerged naturally from the decision to maintain consistent bearing sizes and journal diameters as displacement grew.

The History of the Small Block Chevy Motor

Chevrolet offered both cast and forged variants depending on application, with forged cranks used where higher stress was expected. Cast cranks worked fine for passenger car duty; they were less expensive to produce and perfectly adequate for engines that would rarely see sustained high RPM. Performance applications received forged cranks with superior material properties and fatigue resistance. This tiered approach lets Chevrolet optimize cost and performance for each specific use case.

Connecting rods and pistons were sized to manage the expected RPM and power levels while keeping reciprocating mass under control. As performance demands increased, the basic geometry supported upgrades such as forged rods, lighter pistons, and improved ring packages without requiring a redesign of the entire bottom end. The original dimensions and bearing sizes provided sufficient margin to accommodate significantly higher loads than originally specified.

A production 350 Small Block used in a full-size sedan might spend its life below 4,500 RPM, yet that same crankshaft architecture has been safely turned far higher in performance builds with upgraded components. The five-main-bearing design provides enough support for 7,000+ RPM operation when paired with appropriate balancing, oil control, and valvetrain stability.

Stroker combinations, such as 383-cubic-inch builds using a longer-stroke crank in a 350 block, demonstrate how flexible the original journal layout and clearances are. The architecture welcomes creative reconfiguration. Want more displacement? Install a longer-stroke crank. Want higher RPM? Use lighter pistons and better rods. The basic platform accommodates both approaches without fundamental redesign.

When you think about increasing power, do you automatically consider what it means for the rotating assembly, bearing loads, piston speeds, and balance? Many enthusiasts focus on airflow and camshaft timing without recognizing that the bottom end must survive the forces their modifications create. Power isn't just about getting air and fuel into the cylinders; it's about managing the tremendous mechanical loads generated when that mixture ignites.

How might Chevrolet's choice to standardize certain dimensions across displacements have unintentionally laid the groundwork for the stroker combinations so popular today? By maintaining consistent bore spacing, deck height, and journal sizes, they created a system where components could be mixed and matched far beyond their original applications. That flexibility became one of the Small Block's most valuable characteristics.

Section 2.4: Cylinder Heads and Combustion Chambers – Efficient Power from Simple Shapes

The Small Block's cylinder heads, particularly their wedge-shaped combustion chambers and valve layout, strike a careful balance between manufacturability, efficiency, and flow potential. They are central to the engine's response to both factory and aftermarket development.

From the beginning, Chevrolet designed the heads to promote good mixture motion and stable combustion while staying simple to cast and machine. The wedge-shaped combustion chambers encourage swirl and efficient flame travel across the bore. This shape, wider at one end than the other, creates turbulence that helps mix fuel and air while providing a relatively direct path for the flame front to travel from the spark plug across the piston crown. It's not as sophisticated as modern pent-roof or hemispherical designs, but it's far more practical to manufacture and delivers surprisingly good efficiency.

The in-line valve arrangement simplified rocker geometry and head casting while still allowing decent port size and shape. Both intake and exhaust valves sit in a straight line along the cylinder, so the intake port opens on one side and the exhaust port on the other. This crossflow layout, with intake on one side and exhaust on the other, improved breathing and thermal management compared to older "siamesed" port designs, where intake and exhaust shared common passages or ran close together.

Critically, the basic architecture provided room for growth. The design supported larger valves, reworked ports, and varied chamber volumes as performance and emissions targets evolved. While early heads had modest airflow, the underlying geometry allowed engineers and aftermarket porters to steadily increase airflow without abandoning the core design. Over time, this made the Small Block a moving target, in a good way, for incremental improvement.

The progression from early "Power Pack" heads to later high-performance castings such as "fuelie" heads showcases how Chevrolet extracted more power from essentially the same architecture by altering port shapes, valve sizes, and chamber volumes. Each generation refined what came before, but the fundamental wedge-chamber, inline-valve layout remained constant. That consistency meant lessons learned on one casting applied to others, knowledge accumulated rather than reset with each new design.

Modern aluminum replacement heads often bolt onto a 1960s short-block but feature optimized ports and refined chamber designs. This backwards compatibility is only possible because the original layout was so fundamentally sound. The bolt patterns, port locations, and basic geometry were correct from the beginning, which meant future improvements could build on that foundation rather than working around its limitations.

Do you tend to think of cylinder heads as static "parts," or as the primary tuning instrument for how an engine breathes and burns fuel? The truth is that heads arguably matter more than any other single component in determining an engine's power characteristics. The block provides displacement and structural support, but the heads control how efficiently that displacement can be filled, ignited, and emptied. Port shape, valve size, chamber volume, and valve angle all profoundly affect volumetric efficiency, compression ratio, and power delivery.

How does knowing the Small Block's basic head architecture help you understand why some builds focus heavily on head selection and port work? Because once you recognize that the heads are the gatekeepers of airflow, it becomes obvious why a mediocre set of heads will limit even an otherwise excellent engine, and why great heads can transform a modest combination into something genuinely impressive.

Section 2.5: The Lubrication System – Lifeblood Under Pressure

The Small Block's lubrication system, oil pump, galleries, and distribution paths were engineered for reliable, even oil delivery across a wide range of operating conditions. Its straightforward layout is a major reason these engines survive abuse that would quickly kill more fragile designs.

Oil in the Small Block performs multiple roles: reducing friction between moving parts, carrying heat away from highly stressed areas such as bearings and piston rings, and cleaning internal surfaces by suspending contaminants until they can be filtered out. The system that manages this is deceptively simple in concept but carefully engineered in execution.

A gear-driven oil pump located in the oil pan is driven directly off the distributor shaft via the camshaft. This mechanical drive means oil pressure rises immediately with engine speed, with no belts to slip and no external drives to fail. The pump draws oil from the pan through a pickup tube, pressurizes it, and sends it through the oil filter before distributing it throughout the engine.

The main oil gallery layout feeds main bearings, cam bearings, and lifters in a logical sequence. Oil enters the main gallery, typically a drilled passage running the length of the block, then branches off to each main bearing. From the mains, oil flows through drilled passages in the crankshaft to the rod bearings. Simultaneously, passages in the block feed the camshaft bearings and lifter bores.

The History of the Small Block Chevy Motor

Pushrod oiling to the top end represents an elegant solution to valvetrain lubrication. Oil flows up through hollow pushrods to lubricate rocker arms and valve tips. This design eliminates the need for separate oil lines or external plumbing to the cylinder heads; the pushrods themselves serve as oil passages. It's simple, reliable, and difficult to mess up during assembly.

Pressure regulation via an internal relief valve prevents overpressurizing the system at high RPM or during cold starts, when the oil is thick. The relief valve opens when pressure exceeds its setpoint, dumping excess flow back to the pan. This protects seals and gaskets while ensuring that oil volume isn't wasted pumping against excessive resistance.

Because the galleries are accessible and well-documented, builders have been able to modify oiling strategies for racing, restricting flow to certain areas, enlarging passages, or using higher-volume pumps, without reinventing the entire system. Priority oiling systems, which ensure the main and rod bearings receive oil before the lifter valley, have become common in high-performance builds. These modifications work because they're refining a fundamentally sound design, not compensating for fundamental flaws.

Stock Small Blocks used in trucks and work vehicles often live long lives under sustained load and less-than-ideal maintenance. Their ability to keep bearings and valvetrain surfaces lubricated even when oil changes are neglected speaks to the design's inherent robustness. The system has sufficient margin, flow capacity, gallery size, and pressure to tolerate imperfect conditions without immediate failure.

Race engines frequently adopt simple but effective lubrication upgrades, windage trays to control oil in the pan, baffled pans to prevent starvation during hard cornering, and improved pickup placement to ensure a constant oil supply, while still relying on the same fundamental pump and gallery layout created in the 1950s. The fact that these relatively minor additions are sufficient to support 9,000+ RPM operation proves the core system's quality.

When you think about performance, how often do you consider oil control and lubrication as part of the package, rather than just horsepower and airflow? Many enthusiasts overlook oiling until something fails: a spun bearing, a wiped cam lobe, or a seized lifter. But proactive attention to lubrication prevents those failures and extends engine life.

What does the Small Block's reputation for durability tell you about the importance of a well-engineered lubrication system, even if most owners never see it in operation? It suggests that invisible engineering, the passages, clearances, and flow paths that operate out of sight, matters just as much as the visible components. An engine might make impressive power on a dyno, but if it can't sustain that power for hours or years, the performance is ultimately meaningless.

If you're planning a build that will see extreme use, note one or two lubrication-related upgrades to research further: a baffled oil pan, a windage tray, or simply a high-quality pump and pickup matched to your application. These aren't attractive modifications; nobody sees your oil pan at a car show, but they're the difference between an engine that performs once and one that performs reliably over the long term.

Section 2.6: Cooling System Fundamentals – Controlling Heat for Longevity

The Small Block's cooling system, water jackets, passages, and flow paths manage thermal loads so the engine can operate safely across everything from slow city traffic to highway cruising and racing. Its effectiveness is a quiet but essential reason for the engine's long life and reliability.

Combustion generates intense localized heat, especially around exhaust valve seats and cylinder walls, where hot gases escape at high velocity. The Small Block's casting design addresses this with well-distributed water jackets around cylinders and combustion chambers to even out temperature differences. Rather than allowing hot spots to develop, the jacket design encourages coolant flow near the areas that need it most.

A front-mounted water pump circulates coolant through the block and heads before returning it to the radiator. The pump, typically driven by a belt from the crankshaft, creates flow through a deliberate path: from the radiator into the block, up through the cylinder heads where temperatures are highest, and back to the radiator where heat is rejected to the atmosphere. This continuous circulation prevents localized overheating even when ambient temperatures are high or the engine is working hard.

Thermostat-controlled flow brings the engine up to operating temperature quickly and maintains it consistently. When the engine is cold, the thermostat remains closed, blocking flow to the radiator and allowing coolant to circulate only within the engine. This accelerates warm-up, improving fuel economy and reducing wear during the critical cold-start period. Once coolant temperature reaches the thermostat's setpoint, typically around 180-195°F, the thermostat opens, allowing coolant to flow to the radiator. As temperatures fluctuate, the thermostat modulates flow to maintain a stable operating temperature.

The passage design balances coolant flow between cylinder banks, reducing hot spots that might otherwise shorten engine life. While the Small Block isn't perfect, early designs could develop hot spots in certain areas, particularly around cylinders that received less flow. The basic architecture proved adequate for most applications and left room for refinement in later generations.

Later generations, such as LT1 and LS engines, refined cooling strategies with reverse-flow cooling and improved water-jacket designs, but the original Small Block layout proved more than adequate for the power levels and duty cycles of its era. It also left room for improved radiators, pumps, and coolant management in performance use. Enthusiasts routinely upgrade to larger radiators, higher-flow water pumps, and improved fan setups, modifications that the basic system accommodates without fundamental redesign.

The History of the Small Block Chevy Motor

Many classic Chevrolet cars and trucks still on the road today run their original Small Blocks with upgraded radiators and fans. The fact that these engines tolerate decades of thermal cycling, cold starts on winter mornings, hot-soaked shutdowns after summer highway runs, countless heating and cooling cycles, speaks to effective cooling at the core design level. The castings don't crack, head gaskets don't fail prematurely, and cylinder walls don't distort, because the cooling system maintains reasonably even temperatures under varied conditions.

In racing or towing applications, builders typically retain the basic coolant flow paths while improving heat rejection with larger radiators, high-flow pumps, and better airflow management, again leveraging a solid base design rather than compensating for fundamental flaws. The modifications enhance capability without replacing the underlying system.

Do you tend to think of overheating problems as "radiator issues," or do you consider the entire system, from internal water jackets to airflow through the grille? Many cooling problems have nothing to do with the radiator itself. Insufficient airflow, a weak water pump, a stuck thermostat, air pockets in the system, or blocked passages can all cause overheating even with a perfectly adequate radiator. Effective cooling requires every element to work correctly.

How might understanding the Small Block's cooling pathways influence your choices in thermostats, pump types, or radiator sizing on a future project? Once you recognize that cooling is a system, not just a single component, you'll evaluate upgrades differently. A high-flow water pump only helps if the rest of the system can handle increased flow. A larger radiator only works if adequate air passes through it. These components must work together harmoniously.

If you currently own a Small Block-powered vehicle, note your typical operating temperature, the conditions under which temperature climbs (slow traffic, towing, high ambient heat), and whether your cooling system maintains stability or struggles in certain situations. This data will be useful when you later consider whether cooling-system upgrades are worth the investment, and which specific upgrades will address your actual needs rather than imagined problems.

In this chapter, you've seen how the Small Block Chevy's anatomy is anything but accidental. The thin-wall, deep-skirt block creates a strong yet compact backbone. The pushrod valvetrain offers efficient breathing and easy tuning. The rotating assembly delivers smooth, durable power. Cylinder heads and wedge-shaped combustion chambers enable efficient combustion and scalable performance. Finally, robust lubrication and cooling systems quietly sustain all of this under real-world conditions.

The guiding question, what makes this engine's design so effective, comes down to integration. Each subsystem is competent on its own, but it's the way they work together that makes the Small Block legendary. The block provides structural integrity that supports higher power levels. The valvetrain breathes efficiently while remaining simple to service.

The rotating assembly turns smoothly while accepting a wide range of configurations. The heads flow well while maintaining compact dimensions. The lubrication system protects components under varied conditions. The cooling system manages heat without exotic solutions. No single element is revolutionary, but the combination is extraordinary.

You've also seen a recurring pattern: choices made for manufacturability, reliability, and simplicity in the 1950s ended up creating an engine that could be refined, modified, and pushed far beyond its original brief. That adaptability is the thread that runs through this entire book. The Small Block succeeded not because it was perfect from the beginning, but because it was fundamentally sound and remarkably flexible.

Now that you understand the core anatomy, the next logical question is: how did this basic architecture evolve? In Chapter 3, we'll trace the Small Block's generational journey, from the original 265 through LT1, LS, and modern Gen V designs. You'll see which features remained sacred, which changed radically, and how each generation responded to new performance, efficiency, and emissions while preserving the engine's core identity.

The next time you walk past a Small Block on a stand or hear one fire up at a car meet, take a moment to think about what's happening inside: the crank turning in its deep-skirt block, pushrods snapping valves open and closed, oil and coolant quietly doing their jobs. Once you see the engine as a carefully orchestrated system rather than a collection of parts, you're not just admiring a classic design; you're beginning to think like the engineers and builders who turned this compact V8 into an enduring legend.

The History of the Small Block Chevy Motor

Chapter 3: Generations of Power – Evolution and Transformation

Visualize removing the valve cover from a 1955 265 and then from a modern LT1. At first glance, they seem to belong to entirely different eras, which they do. However, beneath the aluminum castings, coil packs, sensors, and direct injectors lies a shared architectural DNA that traces directly back to Ed Cole's original vision. The Small Block didn't endure by remaining the same; it thrived by evolving intelligently.

In Chapter 1, you saw how market pressure and bold leadership brought the original Small Block into existence. Chapter 2 opened that engine up on the stand and walked through its anatomy. Now we step back and watch it move through time, generation by generation, as Chevrolet engineers adapted the design for new performance expectations, emissions regulations, and manufacturing technologies.

Understanding this evolution is crucial if you're trying to identify engines, plan a build, or simply appreciate why a 350, an LT1, an LS3, and an LT4 can all be "small-blocks" and yet behave so differently. By the end of this chapter, you'll be able to differentiate the major

Small Block generations and place them in historical context, recognize the key design features and technological innovations that define each generation, understand the basic displacement families within each era and how they were applied in production and performance vehicles, and make more informed decisions about parts interchangeability, upgrade paths, and engine selection for your own projects.

This chapter is organized as a chronological roadmap. Each section focuses on a major generational phase in the Small Block's development, followed by a synthesis section that ties patterns together for enthusiasts and builders.

Section 3.1: Generation I – The Foundational Era (1955–2003)

Generation I Small Blocks established the blueprint: compact, lightweight, pushrod V8s that could be cast in huge volumes, tuned for wildly different applications, and continuously refined without losing their basic architecture. Think about that for a moment, almost five decades of production with the same fundamental design. That's not just longevity; that's proof of concept on a scale few engines ever achieve.

Gen I covers an astonishing span that begins with the 265 in 1955 and runs through the 283, 327, 350, and 400 cubic-inch variants, among many others. Throughout this nearly half-century journey, core characteristics remained stable: a 90-degree V8 configuration with a single camshaft in the block (that cam-in-block or OHV design that became the Small Block's signature), two-valve wedge-chamber cylinder heads that prioritized flow without complexity, 5-bolt and later 4-bolt main bearing variations for different strength needs, and carburetion initially, later transitioning to throttle-body and port fuel injection on certain models.

Over time, Chevrolet increased displacement primarily by changing the bore and stroke, a strategy that allowed it to extract more power without redesigning the entire architecture. The 265 and 283 emphasized compact size and revving capability, appealing to enthusiasts who wanted an engine that could spin freely and deliver its power at high rpm. The 327 and 350 balanced street torque and high-RPM performance, creating what many consider the sweet spot of the Gen I family, engines that worked equally well in daily drivers and weekend warriors. The 400 pushed displacement boundaries to meet rising power demands in heavier vehicles and performance applications, though it came with compromises like siamesed cylinders, which limited cooling capacity.

Throughout this era, emissions regulations in the 1970s and 1980s forced compression drops, more conservative camshafts, and the introduction of catalytic converters and electronic controls. These weren't changes engineers wanted to make; they were changes they had to make to keep the Small Block legal and relevant. Yet even with these restrictions, the underlying block and head patterns remained recognizable. A mechanic who learned on a 1955 265 could still work on a 1985 305, even if the specifics had changed.

Consider the 283 "fuelie" from 1957, which showcased mechanical fuel injection and achieved the famous one-horsepower-per-cubic-inch benchmark, a figure that seemed almost magical at the time. Or look at the 327 in mid-1960s Corvettes and Camaros, which became a favorite among performance enthusiasts for its rev-happy nature and willingness to make power anywhere in the RPM range. The 350, introduced in the late 1960s, eventually became the default small-block; the engine swapped into everything from street rods to pickup trucks because of its balance of parts availability and performance potential. And the 400, with its identical cylinders and external balance, demonstrated how far Chevrolet could stretch the original block before moving to a new generation.

When you hear "Small Block Chevy," which displacement comes to mind first,265, 283, 327, 350, or 400, and why do you think that is? Most enthusiasts default to the 350, and there's a reason: it represents the perfect intersection of availability, capability, and upgrade potential. How do you see the trade-offs Chevrolet made in Gen I, between emissions, fuel economy, and performance, showing up in the engines you've worked on or driven? Those 1970s and early

1980s engines, in particular, show the scars of compromise. Lower compression, restrictive exhaust systems, and conservative camshafts robbed power in the name of emissions compliance, but they also created a foundation that responds beautifully to modern performance upgrades.

If you have access to a Gen I engine, or even detailed photos, take a moment to identify the casting numbers on the block and heads, then determine the displacement and approximate production era they correspond to. Use an online casting database or reference book to decode them. Note how many years and vehicle types that specific casting served. This exercise drives home just how universal the Gen I architecture became. A single casting might appear in everything from a Corvette to a pickup truck, from a Camaro to a station wagon. That's not corner-cutting; that's intelligent engineering that maximizes production efficiency while maintaining flexibility.

Section 3.2: The LT1/LT4 Era – Bridging Old and New (1992–1997)

The early 1990s LT1 and LT4 engines represent a pivotal "bridge" moment: they retain Gen I block lineage while integrating modern technologies such as reverse-flow cooling and advanced ignition to meet higher performance and stricter emissions standards. If Gen I was the foundation, the LT1/LT4 era was the moment Chevrolet asked itself a critical question: Could we modernize the Small Block without abandoning what made it great in the first place?

Often called "Gen II" by enthusiasts (though Chevrolet never officially designated them that way), these engines evolved the traditional small-block rather than replacing it outright. Key innovations included reverse-flow cooling, where coolant flowed to the heads first, allowing higher compression ratios without detonation, a clever solution that addressed one of the Gen I's thermal management challenges.

The Opti-Spark distributor, mounted at the front of the crank, improved ignition accuracy by firing directly off the crankshaft position rather than through a traditional distributor drive. However, it added complexity and some reliability concerns when moisture entered the system. Higher-flow cylinder heads and camshaft profiles tuned for both low-end torque and higher RPM power gave these engines a broader, more usable power band. And multi-port fuel injection, combined with more sophisticated engine management systems, enabled precise control of fuel delivery and ignition timing across the entire operating range.

The LT1 and LT4 were designed to deliver performance to cars like the Corvette and Camaro that could compete globally while still being based on a familiar manufacturing footprint. This was crucial; Chevrolet needed engines that could be built on existing production lines without massive retooling investments.

The 300-plus-horsepower LT1 in the C4 Corvette and fourth-gen Camaro/Firebird set new expectations for out-of-the-box performance in the early 1990s. These were engines that felt genuinely quick without modification, delivering power with a smoothness and refinement that earlier Gen I engines rarely matched.

The LT4, offered in the 1996 Corvette Grand Sport and certain special models, pushed output further with better heads, cam, and valvetrain components, essentially showing what the LT1 architecture could do when optimized for maximum performance rather than balancing cost and capability. In the B-body Impala SS, the LT1 demonstrated that the same architecture could deliver strong, refined torque in a full-size sedan, proving the engine's versatility extended beyond sports cars.

When you look at the LT1/LT4, do you see them more as "modernized Gen I" engines or as the first step toward the LS revolution? What specific features shape your view? There's no single correct answer here; these engines genuinely straddle two worlds. They retain the external dimensions and mounting patterns of Gen I, making them relatively straightforward swaps into older vehicles.

Yet, they incorporate technologies that pointed directly toward what was coming next. How might reverse-flow cooling and tighter electronic control have changed the way enthusiasts tune these engines compared to an older carbureted 350? The answer reveals a fundamental shift: tuning moved from mechanical adjustments, jetting carburetors, recurving distributors, and swapping springs, to software modifications and electronic calibration.

Sketch a quick comparison table with three columns: Gen I traditional (say, a 1980s 350), LT1/LT4, and early LS1. List cylinder head material and design features, fuel delivery method, ignition type, and cooling strategy. This simple exercise helps you visualize the evolutionary steps that lead from the original block to the all-new LS family. You'll see how each generation built on its predecessor's strengths while addressing specific weaknesses or limitations.

Section 3.4: The LS Generation – A Modern Small Block Revolution (1997–2013)

The LS family represents a clean-sheet reinterpretation of the Small Block ethos, retaining cam-in-block pushrod simplicity while embracing aluminum construction, optimized port geometry, and advanced engine management to deliver huge gains in power, efficiency, and durability. If you're an enthusiast who came of age after the turn of the millennium, the LS is probably what you think of when someone says "Small Block Chevy." It's that influential.

Introduced with the LS1 in the 1997 Corvette, this generation is mechanically distinct from Gen I, but philosophically aligned. It's still a 90-degree, OHV V8 with two valves per cylinder. It's still compact and relatively lightweight for its output. It's still designed for mass production and wide application, from sports cars to trucks. But the details changed substantially, and those details made all the difference.

Key LS innovations included aluminum (and some iron) blocks with a deep-skirt design and six-bolt main bearing caps to increase bottom-end rigidity. This structural improvement allowed the engine to handle significantly more power without requiring exotic materials or extensive reinforcement. Coil-near-plug ignition eliminated the traditional distributor, improving spark accuracy and reliability while reducing complexity; there are no more distributor gears to wear, no timing chains driving secondary components.

Cathedral-port (early) and later rectangular-port heads featured highly efficient airflow and combustion characteristics, delivering impressive power figures with relatively mild camshafts and conservative compression ratios. Improved sealing, oiling, and block architecture supported high RPM and high power levels with minimal modification. Many LS engines can safely spin past 7,000 RPM with nothing more than appropriate valvetrain components. Integration with advanced engine control modules (ECMs) allowed precise fuel and spark management that would have been impossible just a generation earlier.

The LS family grew into a full ecosystem: LS1, LS6, LS2, LS3, LQ4/LQ9 (iron truck variants), and high-performance supercharged engines like the LS9. Each variant served specific applications and price points, but they shared enough common architecture that parts interchangeability remained high within the family.

The LS1 in the C5 Corvette proved that a light, efficient pushrod V8 could compete head-to-head with more complex multi-cam imports, often beating them in both power-to-weight ratio and reliability. The LS3 and L92-style heads became favorites in both factory cars and the swap world due to their excellent flow and reasonable cost. These heads, with their rectangular intake ports and improved combustion chamber design, represented a significant step forward in breathing efficiency. Truck-based LS engines,5.3L and 6.0L iron blocks, formed the backbone of the budget performance movement. They're readily available in junkyards, incredibly durable, and highly responsive to cam and head upgrades. A $500 junkyard 5.3L with a cam swap and better heads can easily produce 450+ horsepower, making it one of the best dollar-per-horsepower propositions in performance engine building.

Why do you think the LS quickly became the new "default swap" engine, displacing the traditional 350 in many builds? Which features of the design make it particularly attractive? Consider the combination of factors: lighter weight, more power in stock form, better efficiency, coil-near-plug reliability, and widespread availability.

When you can pull a complete LS engine from a wrecked truck for less than rebuilding a worn-out Gen I small-block, the economics become compelling. Look at how the LS maintained a pushrod layout in an era dominated by overhead cam designs. What does that tell you about GM's priorities and the strengths of the Small Block concept? It suggests that complexity isn't always the path to better performance; sometimes, refining a proven architecture delivers superior results.

Choose one LS variant, LS1, LS3, or a common 5.3L truck engine, and research its factory horsepower and torque ratings, typical real-world power numbers with basic bolt-ons (intake, exhaust, tune), and common swap platforms where this engine is used. Note how often the same long-block architecture appears in radically different vehicles. This is the Small Block adaptability principle, updated for the modern era. You'll find LS engines in everything from drag cars to off-road trucks, from vintage hot rods to modern sports cars, proving that the core concept remains as versatile as ever.

Section 3.5: Generation V – Direct Injection and the New Efficiency Frontier (2014–Present)

Generation V (often called LT1/LT4/LT5 in modern form) brings the Small Block into the world of direct injection, variable valve timing, and sophisticated fuel-saving strategies, meeting 21st-century emissions and efficiency standards while preserving V8 character and performance. If the LS generation proved the Small Block could compete with modern multi-cam engines, Gen V proved it could meet regulations that seemed impossible just a decade earlier.

Gen V builds on the LS-based architecture but introduces significant changes. Direct injection (DI) delivers fuel directly into the combustion chamber rather than through the intake port, enabling higher compression ratios and more precise control over combustion characteristics. This technology, common in modern engines, lets engineers run compression ratios that would cause detonation with port injection, extracting more power and efficiency from every drop of fuel.

Variable valve timing (VVT) alters camshaft phasing on the fly to improve low-end torque, high-RPM power, and fuel economy, essentially allowing a single camshaft to behave like several different cams depending on operating conditions. Active Fuel Management (AFM) or cylinder deactivation shuts off cylinders under light load to reduce fuel consumption, a controversial feature among enthusiasts but one that allows V8s to survive in an increasingly efficiency-focused regulatory environment. Revised combustion chambers, intake ports, and piston crowns optimized for DI and emissions performance complete the package.

Despite these complexities, the core remains: cam-in-block, two-valve-per-cylinder OHV layout, compact external dimensions compared to many overhead-cam competitors, and high specific output in performance trims, such as supercharged LT4 and LT5 variants. Generation V shows how far the small-block idea can be pushed without abandoning its essential mechanical identity.

The LT1 in the C7 Corvette and late-model Camaro SS demonstrates how a naturally aspirated Gen V can offer both strong performance and respectable fuel economy. These engines produce impressive power numbers, around 455-460 horsepower in Corvette trim, while returning surprisingly good fuel economy when driven conservatively. The supercharged LT4 in the Corvette Z06 and Camaro ZL1 delivers supercar-level power (around 650 horsepower) from a compact OHV package, proving that pushrod engines can still compete at the highest performance levels. Truck and SUV variants use AFM and VVT to manage the demands of towing, efficiency, and emissions without giving up the torque customers expect, a balancing act that would have been impossible without sophisticated electronic control.

The History of the Small Block Chevy Motor

How do you feel about technologies like direct injection and cylinder deactivation being integrated into a Small Block? Do they enhance or dilute the engine's traditional character in your view? This question divides enthusiasts. Some see these technologies as a necessary evolution, allowing the V8 to survive regulatory pressures that would otherwise force electrification. Others view them as added complexity that undermines the Small Block's traditional simplicity and reliability. Both perspectives have merit. , DI and AFM do add complexity and potential failure points, but they also allow V8 engines to remain viable in markets where they'd otherwise be regulated out of existence.

Looking ahead, what aspects of Gen V design seem most likely to influence future internal combustion engines, even as electrification expands? Direct injection will almost certainly remain standard, as will variable valve timing. Whether cylinder deactivation becomes universal or fades depends on how quickly electrification displaces traditional power plants.

If you're considering a modern swap or late-model performance build, identify one Gen V engine code (LT1, L83, L86) and its factory-rated specs, along with any known challenges when integrating it into older vehicles, fuel system compatibility, electronics, and exhaust routing. Document how these challenges differ from a typical LS swap. This comparison clarifies where the technology leap from LS to Gen V has real-world implications for builders. Gen V swaps require more sophisticated fuel systems (high-pressure DI pumps and low-pressure port injection on some variants), more complex wiring harnesses, and careful attention to exhaust routing around the DI fuel system components.

Section 3.6: Displacement Families, Applications, and Interchangeability

Across all generations, the Small Block evolved through distinct displacement "families" tailored to specific roles, compact, rev-happy engines; balanced street/strip workhorses; and large-displacement torque producers, each with its own strengths and compatibility quirks. Understanding these families is crucial for anyone planning an engine build or trying to identify what they've got on the stand.

Within each generation, Chevrolet offered multiple displacements to match vehicle needs. Gen, I included 265, 283, 302, 305, 307, 327, 350, 400, and others, serving everything from compact coupes to heavy sedans and trucks. This variety allowed Chevrolet to optimize each engine for its intended application rather than forcing a one-size-fits-all approach. The LT1/LT4 era was primarily 5.7L (350 cubic inches), tuned differently for Corvettes, F-bodies, and full-size sedans, showing that displacement is only one factor in engine character.

The LS family exploded with variety: 4.8L, 5.3L, 5.7L, 6.0L, 6.2L, and more, split between performance cars and trucks. This range gave enthusiasts unprecedented choice when selecting an engine for a project. Gen V continues this tradition with 4.3L V6 derivatives (yes, V6, the Small Block architecture proved adaptable enough to shed two cylinders), 5.3L, 6.2L, and high-performance supercharged variants.

Parts interchangeability is both a strength and a potential trap. Many Gen I components swap across displacements, but details like main journal size (early engines used smaller mains, later ones larger for increased strength), balance (internal versus external), and cooling passages (which changed over the years) matter tremendously. LS platforms share a high degree of interchangeability in heads and intakes within subfamilies. Still, port shapes (cathedral versus rectangular), injector styles, and sensor provisions vary enough to create compatibility issues if you're not careful. Gen V parts are less backward-compatible due to DI and revised architecture, but still maintain internal family interchange within that generation.

Understanding which parts carry over and which do not is central to building reliable combinations rather than Frankenstein engines that fight themselves. A classic mistake is mixing components from different eras without understanding the implications, using the wrong flexplate or balancer on an externally balanced engine, for example,

The History of the Small Block Chevy Motor

or trying to bolt Gen I accessories onto an LS block without adapter brackets.

Consider a traditional "383 stroker" Gen I build that combines a 350 block with a 400 crank to increase displacement. This popular modification illustrates how bore-and-stroke mixing can be used within a family to create combinations that Chevrolet never offered from the factory. The result is an engine that fits in 350 engine mounts, uses 350 heads and accessories, but delivers 400-style torque.

Or look at swapping high-flow LS3-style heads onto a 6.0L iron LS truck block to create a budget street/strip engine with significant power gains. The truck block costs a fraction of an aluminum performance block but accepts the same heads, delivering serious performance without the price tag. Contrast this with attempting to mix incompatible components, such as using the wrong flywheel on an externally balanced 400, which demonstrates the importance of understanding balance and application specifics. Get this wrong, and you'll have an engine that vibrates itself to pieces.

Which displacement family best lines up with your goals: high-revving small cubic inches, balanced mid-size, or big-inch torque? Why? Your answer depends on your intended application. Road racing and autocross often favor smaller displacement with high-RPM capability. Street performance typically benefits from mid-size engines (327, 350, LS3) that balance power and drivability. Drag racing and towing applications often require maximum displacement to maximize torque production.

Have you ever run into a compatibility issue, mismatched flexplate, accessory brackets, or heads, that cost you time and money? What could a clearer understanding of engine families have saved you? These lessons tend to be expensive, but they're also memorable. Once you've bought the wrong water pump or discovered your intake manifold doesn't match your heads' port configuration, you tend not to make that mistake again.

The History of the Small Block Chevy Motor

Pick the engine you currently own or dream of building. Identify its generation, displacement, and typical factory applications. List three compatible upgrade paths, head swap, cam upgrade, intake change, that stay within its generation's "comfort zone." This gives you a grounded, reality-based starting point for planning a build that respects the engine's design instead of fighting against it. Building with the architecture rather than against it typically delivers better results with less frustration and expense.

In this chapter, you've followed the Small Block Chevy from its Gen I roots through the LT1/LT4 bridge engines, into the LS revolution, and finally into the highly sophisticated Gen V era. You've seen how each generation introduced specific innovations, reverse-flow cooling, aluminum deep-skirt blocks, coil-near-plug ignition, direct injection, variable valve timing, and AFM, while preserving the core small-block philosophy of compact, efficient power. You've also explored how displacement families and parts interchangeability shaped real-world applications and project planning for builders and enthusiasts.

The Small Block's survival wasn't accidental. Its ability to evolve, mechanically, electronically, and architecturally, while staying true to its original mission, is what turned it from a single-engine program into a multi-generational dynasty. Understanding this evolutionary map gives you a powerful framework: you're no longer just looking at isolated engines, but at chapters in an ongoing engineering story.

Up to this point, we've focused on design and generational change. But none of that matters unless an engine can actually be built, in huge numbers, with consistent quality. In the next chapter, we leave the engineering offices and step onto the factory floor. Chapter 4, "Revolution on the Assembly Line," will show you how Chevrolet turned the Small Block from a promising prototype into a mass-produced powerhouse that reshaped the company's entire vehicle lineup.

The History of the Small Block Chevy Motor

As you move forward, keep this in mind: every time you see a Small Block, whether it's a greasy 350 on a stand, an LS tucked into a drift car, or a Gen V rumbling under the hood of a modern Corvette, you're looking at a living lineage. Those castings and part numbers are more than hardware; they're milestones in a continuous evolution of power, problem-solving, and performance. The engine hasn't just survived across generations, it's thrived, adapting to each new challenge while maintaining the core principles that made it revolutionary in 1955. That's not just engineering excellence; that's automotive immortality.

Chapter 4: Revolution on the Assembly Line

In early 1955, workers on Chevrolet's engine line watched the first production Small Block V8s roll past their stations. The design was fresh, the tooling was new, and the pressure was intense. Dealers were already taking orders, magazine writers were circling for test drives, and management expected this engine not just to succeed, but to redefine Chevrolet's place in the market. There was no safety net. If the line stumbled, the entire V8 gamble could unravel.

In Chapter 3, you saw how the Small Block's architecture lent itself to evolution across generations, how the overhead valve configuration, thin-wall casting, and compact dimensions created a foundation that could grow and adapt. Here, we step back to the moment when that potential would either be realized or lost: the transition from prototype to production. This chapter shows how manufacturing decisions, production targets, and real-world quality control turned a clever design into a durable legend.

By the end of this chapter, you'll understand the scale and urgency of the 1955 production launch and why it was such a high-stakes move for Chevrolet. You'll recognize how early production numbers, market reception, and press coverage validated (or challenged) the Small Block's promise. You'll trace key production milestones, including the fuel-injected 283 and the one-horsepower-per-cubic-inch benchmark that captured the public imagination.

You'll identify the manufacturing innovations that made high-volume, high-precision engine production possible, and see how early performance- and racing-oriented applications grew directly from the assembly line's success. Finally, you'll be able to apply this production history to practical questions: verifying period-correct engines, spotting later swaps or updates, and understanding why certain casting dates matter when you're evaluating a vintage Small Block.

Section 4.1: Launching a New Kind of Chevrolet

The 1955 model year wasn't just the introduction of a new engine; it was a coordinated production and marketing offensive that would define Chevrolet's identity for decades. The company set aggressive manufacturing targets for the Small Block V8, then had to solve the real-world problems of volume, quality, and consistency in record time.

On paper, the engine was ready. Engineering had signed off, dynamometer results looked promising, and durability tests showed the design could withstand the stresses of everyday driving. The next step was far messier, casting thousands of blocks per day, machining critical surfaces to exacting tolerances, assembling rotating assemblies with minimal variation, and doing it all over and over again without letting quality drift.

Chevrolet's leadership understood that this engine had to be available, not just desirable. Marketing had already begun planting the seeds of anticipation in the automotive press and through dealer networks. The new V8 was being positioned as Chevrolet's answer to Ford and Plymouth's growing reputations for performance, a way to transform the brand from dependable but dull to powerful and exciting. That meant setting initial production goals that matched dealer forecasts and marketing promises, numbers that felt audacious given how recently the engine had been finalized.

Making those numbers required reconfiguring foundries and machining lines to handle thin-wall castings and tighter tolerances than the old Stovebolt six ever demanded. The thin-wall casting technique that made the Small Block so light and compact also made it unforgiving during production. Core placement had to be precise, cooling rates carefully managed, and quality checks rigorous. A shift in the sand cores during casting could leave wall thickness irregular, strong enough in some spots, dangerously thin in others.

Then there was the workforce itself. Assembly line workers who'd spent years building the reliable but straightforward inline-six now had to master new procedures and tools. V8 assembly brought different challenges: more parts to track, tighter torque specifications, and a compact engine bay that demanded careful sequencing. Training programs ran concurrently with early production ramp-up, a high-wire act that left little room for error.

Early on, the company faced the usual growing pains. Scrap rates exceeded projections as foundries dialed in their processes. Minor casting flaws, such as porosity, inclusions, and core shift, required constant vigilance and process adjustments. Machining setups needed tweaking as production volumes exposed weaknesses that prototype runs had never revealed. Supply chain bottlenecks emerged as demand ramped up faster than anticipated, leaving assembly plants scrambling for components.

Initial 1955 production targets called for building enough Small Block V8s to power roughly one-third of Chevrolet's passenger car lineup, an ambitious goal that assumed steady demand and smooth production scaling. Reality exceeded even optimistic projections. As word spread about the new engine's performance and drivability, customer preference shifted dramatically toward V8-equipped models. By year's end, Chevrolet had produced over 700,000 Small Block V8s, volumes that strained capacity and forced rapid expansion of foundry and machining operations.

The History of the Small Block Chevy Motor

A snapshot of a typical 1955 engine plant reveals the complexity behind those numbers. Block casting typically happened at dedicated foundries, such as Saginaw Metal Casting Operations in Michigan, where molten iron was poured into sand molds to form the basic engine structure. Once cooled and cleaned, these raw castings traveled to machining facilities where precision equipment cut cylinder bores, surfaced deck heights, and created the precise geometries that would allow interchangeable parts to fit reliably.

From machining, components moved to assembly plants where workers bolted together rotating assemblies, installed valvetrains, and conducted final inspections before the finished engines shipped to vehicle assembly facilities in cities across the country. This distributed production model allowed Chevrolet to scale quickly, but it also demanded tight coordination and consistent quality standards across multiple sites.

When you think of an iconic engine, do you usually visualize the design room or the assembly line? Most of us imagine the moment of inspiration: Ed Cole's team sketching overhead-valve layouts, calculating displacement, debating combustion-chamber shapes. But the Small Block's real achievement happened in these less glamorous spaces, foundries hot enough to make your eyes water, machining floors loud with the whine of cutting tools. In these assembly lines, the same sequence is repeated hundreds of times per shift, knowing the production pressure behind the 1955 launch changes how you view the engine. It wasn't just clever engineering; it was clever engineering that could be reliably replicated at a massive scale under intense time pressure.

If you own or plan to buy a 1955–1956 Small Block–equipped car, this history raises important questions. Is the engine in the car original to that vehicle, or was it replaced during the decades of use? Does the casting date align with the car's production date, or does it suggest a later swap? What modifications might have been made during the engine's life, and which represent period-correct upgrades versus later changes? These aren't just collector concerns; they affect everything from parts availability to authenticity and value.

Section 4.2: Market Reaction and the First Wave of Public Scrutiny

The Small Block's survival depended on more than just running down a production line without drama. Once these engines found their way into customer hands, a different kind of test began. Buyers, dealers, and automotive journalists all had to be convinced that this compact V8 delivered real-world performance, drivability, and reliability, day in and day out.

Once cars hit showrooms in late 1954 and early 1955, Chevrolet shifted from internal testing to external proof. Period road tests, owner feedback, and dealer service reports became an early feedback loop, highlighting both strengths and teething issues. This was the moment when marketing promises met reality, and the Small Block's reputation would either take hold or crumble.

Key factors shaping that early reputation included acceleration and drivability compared to the outgoing six-cylinder engines. The 265 cubic-inch V8 in base form produced 162 horsepower, modest by later standards but a revelation compared to the 136-horsepower "Blue Flame" six it replaced. More impressive than peak horsepower was how the Small Block delivered its power. Mid-range torque came on strong and smooth, making passing maneuvers confident and merging onto highways effortless. The engine didn't need to be revved to redline to feel responsive; it had power available across a broad rpm range, exactly where drivers actually used it.

Fuel economy relative to the newfound power mattered, too. This was an era when V8s were still a novelty to some buyers, associated with thirsty, premium-fuel-drinking luxury cars. Chevrolet's engineers had worked hard to keep the Small Block efficient, and early reports confirmed reasonable fuel consumption that didn't punish buyers at the pump. Real-world owners reported mileage figures that weren't dramatically worse than the old six, a critical selling point that helped overcome hesitation about V8 ownership costs.

The History of the Small Block Chevy Motor

Serviceability became another unexpected advantage. Mechanics who initially approached the Small Block with caution, another new engine meant another learning curve, quickly discovered it was straightforward to work on. The overhead valve design was more complex than the old flathead configurations that some shops still serviced. Still, the parts were logical, the procedures were clear, and the compact dimensions made access easier than with larger V8s. Dealers could quickly adapt to the new configuration without excessive downtime or confusion, keeping customer satisfaction high.

Automotive magazines and newspapers played a crucial role in shaping public perception. *Motor Trend*, *Road & Track*, and regional automotive columnists all ran tests of the new Chevrolet V8, and their verdicts carried weight. Reviews consistently praised the Small Block's smoothness, willingness to rev, and strong mid-range torque. Tom McCahill, the colorful *Mechanix Illustrated* columnist known for his straight-shooting assessments, called the 1955 Chevrolet "the hot setup" and praised the V8's performance. *Motor Trend* ran acceleration tests confirming that Chevy's new V8 could hold its own against, a nd often beat, competing offerings from Ford and Plymouth.

Anecdotal dealer feedback told the same story: increased showroom traffic driven by "V8 curiosity," and test drives that closed sales. Buyers came in skeptical or simply curious, drove a V8-equipped car, and walked out convinced. That test-drive experience, the immediate torque response, the smooth power delivery, the confident acceleration did more to sell the Small Block than any advertisement could.

Early warranty or service campaigns revealed where production or design still needed fine-tuning. Some early blocks showed minor oiling issues under sustained high-rpm operation, traced to oil passage designs that worked fine under normal conditions but struggled when pushed hard. Revised passages and improved baffling addressed the concern. Occasional reports of valve train noise prompted adjustments to rocker arm geometry and pushrod length specifications. Chevrolet responded to these issues quickly, issuing service bulletins and updating production without undermining public confidence. The company treated them as normal refinements, not fundamental flaws, a nd buyers accepted that framing.

How much do you trust period road tests when evaluating classic engines today? It's worth asking yourself whether you see them as marketing, documentation, or both. The truth is, they're a bit of each. Experienced drivers conducted magazine tests and often included measured data, but manufacturers also courted favorable coverage and sometimes provided specially prepared vehicles for testing. Still, the broad consensus across multiple publications and the alignment with owner experiences suggest those early Small Block reviews captured something real.

Think of a modern powertrain you admire. Is your opinion based more on spec sheets, horsepower numbers, torque curves, 0-60 times, or on how owners and reviewers describe living with it? Chances are, it's the latter that really shapes your impression. The same was true in 1955. Advertised horsepower figures might have initially attracted buyers, but they bought the car based on how it felt to drive and whether they trusted it would remain reliable.

Section 4.3: The 1957 Fuel-Injected 283 and the One-HP-Per-Cubic-Inch Milestone

Within just two years of the Small Block's launch, Chevrolet used the existing production architecture to deliver a headline-grabbing achievement: a 283-cubic-inch engine rated at 283 horsepower, thanks to mechanical fuel injection. This wasn't just a performance leap; it was a manufacturing and marketing statement that positioned Chevrolet at the cutting edge of American automotive technology.

The 283 itself represented an evolutionary step from the original 265. Bore increased from 3.75 inches to 3.875 inches while stroke remained at 3.00 inches, adding displacement and breathing capacity without fundamentally changing the engine's character. Improved cylinder head designs, revised valve sizes, and ongoing refinements in machining tolerances created a solid production foundation capable of supporting higher output.

The History of the Small Block Chevy Motor

Adding Rochester mechanical fuel injection introduced challenges that went far beyond bolting on a different induction system. The "Ramjet" fuel injection, as Chevrolet marketed it, used a mechanical pump driven off the camshaft to meter fuel precisely to individual nozzles at each intake port. Getting this system to work reliably in customer cars, not just on carefully prepared test engines, demanded tighter assembly tolerances and careful calibration steps. Each injection system needed individual setup and adjustment, a time-consuming process that slowed production and required skilled technicians.

Quality control became more complex, too. Ensuring the advertised 283 horsepower could be reliably reproduced in customer cars meant verifying not just the engine's mechanical condition but also the fuel injection system's calibration. Chevrolet implemented additional dyno testing for fuel-injected engines, a step that caught issues before engines shipped, but also added cost and time to the production process.

Dealer training represented another hurdle. The fuel injection option was sophisticated, genuinely advanced for 1957, and needed to be sold, serviced, and supported in the field by mechanics whose experience had been with carburetors. Chevrolet developed training programs and service documentation, but many dealers initially struggled with fuel-injection diagnosis and repairing fuel-injection systems. This complexity meant the system worked best for dedicated enthusiasts willing to learn its quirks, while casual buyers often found carbureted engines easier to live with.

Reaching one horsepower per cubic inch became a symbolic achievement that transcended the actual performance numbers. In 1957, this milestone signaled that Chevrolet's mass-production engine line could now deliver performance once associated only with specialized racing engines or exotic European sports cars. It was proof that American engineering could combine accessibility with capability, volume production with precision.

The History of the Small Block Chevy Motor

A look at the 1957 Corvette and full-size Chevy option sheets shows where the fuel-injected 283 fit in the lineup. In the Corvette, it was the top performance option, priced at $484.20, a significant premium over even the high-performance dual four-barrel version. For that money, buyers got 283 horsepower, impressive throttle response, and bragging rights. In full-size passenger cars, fuel injection was rarer still, an exotic option that signaled serious performance intent.

Contemporary advertising heavily featured the 1:1 horsepower-to-displacement milestone. Chevrolet's marketing materials trumpeted "1 HP per cubic inch!" in bold type, often accompanied by images of the distinctive fuel-injection system, with its prominent intake plenum and individual fuel lines. The message was clear: Chevrolet had achieved something special, a technical accomplishment that validated the company's engineering leadership.

Owner and mechanic experiences with early fuel-injected 283s revealed the trade-offs. High performance? Absolutely, the system delivered sharp throttle response and strong power across the rpm range. But tuning and service complexity separated casual drivers from dedicated enthusiasts. Fuel injection didn't tolerate neglect the way carburetors often did. Clogged injectors, air leaks, and calibration drift all created drivability issues that required knowledge and patience to diagnose. For owners willing to learn the system, fuel injection was rewarding. For those who just wanted to drive without thinking about fuel delivery, a well-tuned carburetor was simpler.

When you see a milestone figure, like "one horsepower per cubic inch", what do you think about? Do you focus on the engineering that made it possible: improved breathing, precise fuel metering, efficient combustion? Or do you consider the manufacturing discipline required to build thousands of engines that actually meet that claim, not just in ideal conditions but across production lots, assembly plants, and real-world service conditions? Both matter, and neither works without the other. The Small Block's achievement wasn't just

The History of the Small Block Chevy Motor

reaching the milestone; it was making that performance accessible and repeatable.

If you're considering a period-correct fuel-injected build today, how much complexity are you willing to accept for authenticity's sake? Modern electronic fuel injection offers better drivability, easier tuning, and superior reliability than 1957's mechanical system. But a Rochester fuel-injected 283 is historically significant in a way that a modern aftermarket EFI system isn't. It's a choice between convenience and originality, a nd there's no wrong answer, just trade-offs you should understand before committing.

Make a simple comparison chart: list a standard carbureted 283 and the fuel-injected 283 side by side. Note factory horsepower ratings, the base 283 ranged from 185 horsepower with a single two-barrel carburetor up to 270 horsepower with dual four-barrels, while fuel injection hit 283. List intended applications, Corvette versus passenger car availability. Record what additional components and setup the injection system required: fuel pump, distribution block, individual injectors, and calibration procedures. This exercise helps you see the production leap from "high volume" to "high volume with high precision," and appreciate why fuel-injected cars commanded premiums then and remain highly collectible now.

Section 4.4: Hitting Milestones: Scaling to the Million-Engine Mark

One of the clearest signs that Chevrolet had solved the production puzzle was sheer volume. With hundreds of thousands of engines built by the end of 1955 and the one-millionth Small Block following soon after, the engine transitioned from "new idea" to "industry fixture", a foundational product that would define Chevrolet's capabilities for decades.

High volume isn't just a bragging point; it reshapes everything about how an engine exists in the world. Parts economies of scale make components more affordable, both for Chevrolet's own production lines and, eventually, for the aftermarket that would grow

The History of the Small Block Chevy Motor

up around the Small Block. When you're casting hundreds of thousands of blocks per year, the per-unit cost of tooling, foundry setup, and machining infrastructure spreads across massive production runs. That efficiency benefits everyone: Chevrolet could price V8-equipped cars competitively, and replacement parts remained affordable for owners.

Interchangeability improved as machining and casting processes stabilized around proven standards. Early production always involves some variation as procedures get refined, but by the time Chevrolet hit the million-engine milestone, tolerances had tightened,d and procedures had been documented thoroughly. An engine built in Tonawanda, New York, could use components manufactured in Saginaw, Michigan, with confidence that they'd fit and function correctly. That standardization simplified service and made the Small Block genuinely versatile across applications.

Multiple vehicle lines, sedans, wagons, sports cars, and eventually light trucks, could share a common engine family, simplifying logistics and service. This was strategic brilliance. Instead of maintaining separate engine families for different vehicle segments, Chevrolet could tune a single basic architecture for various roles. A low-compression 283 for taxi duty, a high-performance 283 for Corvettes, a truck-spec 283 with different intake and exhaust, all shared core components and manufacturing processes. Dealers stocked one set of parts patterns; mechanics learned one basic engine layout.

Reaching the million-engine milestone also signaled internal confidence. Management could justify continued investment in tooling upgrades, capacity expansion, and incremental improvements because the market had clearly embraced the Small Block. This wasn't a risky experiment anymore; it was a proven platform worth nurturing and developing.

For enthusiasts today, this era explains why early Small Block parts patterns and design cues became the "default" architecture for American V8 performance. When you have millions of engines in

The History of the Small Block Chevy Motor

service, the aftermarket follows. Speed equipment manufacturers, machine shops, and rebuilders all oriented their businesses around the most common platform. That created a self-reinforcing cycle: the Small Block was popular, so aftermarket support was strong, which made it more attractive for builds and swaps, which kept it popular.

A timeline of production growth shows the ramp: by late 1955, Chevrolet was building Small Blocks at a rate approaching 60,000 per month. By 1957, monthly production sometimes exceeded 75,000 units. The millionth engine rolled off the line in early 1957, barely two years after its introduction. That pace of scaling is remarkable even by modern standards, a testament to how quickly Chevrolet could expand foundry capacity, train workers, and coordinate supply chains.

Examples of the same core engine family appearing in multiple models show the versatility Chevrolet achieved. A 1955 Bel Air sedan might have a 265 tuned for smooth, efficient family transportation. A 1956 Corvette received the same basic 265 but with revised heads, camshaft, and carburetion for significantly higher output. By 1957, the Nomad wagon, Biscayne taxi special, Two-Ten sedan, and Corvette roadster all drew from the same Small Block family, each configured for its specific duty cycle: family transport, commercial durability, sporty performance, all from the same production architecture.

Period press or internal GM recognition of the million-engine milestone treated it as more than a number. It represented an inflection point in corporate identity and market positioning. Chevrolet was no longer the also-ran in performance, trailing Ford or Plymouth. The Small Block had transformed the brand's reputation, and production volumes proved customers believed the transformation was real.

When you see how many Small Blocks were built, does it change how you think about rarity and value? With millions produced, what makes a particular engine or configuration special? The answer lies in specifics: unusual option combinations, particularly early or late production dates, documented racing history, and matching-numbers status with desirable vehicles. Volume production doesn't eliminate collectibility; it just shifts what collectors value toward the genuinely unusual within the common platform.

For your own projects, do you tend to trust high-volume, well-proven parts? There's wisdom in following where millions of engines have gone before. When a design has been tested across decades and countless miles, you can have confidence in what works and what doesn't. That doesn't mean unthinkingly assuming every factory part is perfect; we'll discuss verification shortly, but it does mean the basic architecture has proven itself durable.

If you're evaluating an engine for a build, identify where its casting date and configuration fall within the broader production timeline. Is it from the early, rapidly evolving years, 1955 through 1957, when changes came frequently, and certain parts became obsolete quickly? Or is it from a later, highly standardized phase when year-to-year differences were minimal? That context influences your expectations for originality, parts availability, and value. An early, rare configuration might be historically significant but challenging to restore correctly. A later, common version might be less exotic but far easier to support with readily available parts.

Section 4.5: Manufacturing Innovations: Precision at High Volume

The Small Block's success depended on more than clever design principles like overhead valves and compact dimensions. It required a parallel revolution in manufacturing,thin-wall casting control, improved machining methods, and structured quality control that allowed Chevrolet to build an advanced V8 to tight tolerances, day after day, on a mass-production line.

Key production advances made that possible. Thin-wall casting control represented a significant technical challenge. The foundries had to precisely manage mold design, core placement, and cooling patterns to maintain block strength without excess weight. Too thick, and you lose the weight advantage that made the Small Block competitive. Too thin, and you risk weak spots, porosity, or outright structural failures. The sweet spot required careful attention to sand core positioning, slight shifts during the casting process could leave one cylinder wall thinner than spec, and controlled cooling rates to minimize internal stresses.

Standardized machining sequences ensured critical features aligned correctly. Crankshaft journals required precise diameters and surface finishes to support bearings properly. Cam tunnels required accurate positioning to maintain valve timing across the rpm range. Deck surfaces had to be flat and parallel to cylinder bores to seal head gaskets reliably. Lifter bores demanded consistent diameter and spacing. Each of these features depended on machining setups that could hold tolerances across thousands of parts without drifting out of spec. Chevrolet invested in dedicated transfer lines, automated systems that moved blocks through sequential machining operations, and implemented statistical process controls that caught problems early.

Assembly-line torque and fitment procedures refined repeatability, helping prevent leaks and premature failures. Bolting patterns followed specific sequences to distribute clamping loads evenly. Torque specifications were carefully chosen to ensure gaskets sealed without crushing or distorting mating surfaces. Gasket technologies improved, too, with better materials and designs that tolerated minor imperfections while still sealing reliably. These seemingly minor details, repeated hundreds of thousands of times, determined whether the Small Block earned a reputation for reliability or became known for frustrating oil leaks and cooling problems.

In-line inspection and testing kept process drift in check as volumes increased. Spot checks caught issues before they cascaded into larger problems. Sample engine teardowns revealed whether internal components showed unexpected wear patterns that might indicate assembly or design concerns. Periodic dynamometer testing verified that power output remained within specification across production lots. This continuous monitoring created feedback loops that allowed Chevrolet to refine processes without halting production, adjusting a foundry's cooling rate here, tweaking a machining setup there.

These manufacturing practices didn't just serve Chevrolet in the 1950s; they laid the groundwork for decades of parts interchange and made the Small Block a natural canvas for future modifications. Because production processes stabilized early and remained relatively consistent, parts patterns from different years and plants often swapped successfully. That flexibility is part of what makes the Small Block so attractive for builds: you can mix and match components from various generations with reasonable confidence they'll work together.

A simplified walk-through of the block's journey shows the discipline involved. Raw casting begins when molten iron is poured into sand molds formed by carefully positioned cores. As the iron cools, it solidifies into the block's basic shape, cylinder bores, water jackets, mounting bosses, and oil passages. Once cooled, the rough casting gets cleaned, sand cores are removed, casting flash is ground away, and the piece is inspected for obvious flaws. From there, it moves to machining, where precision equipment performs dozens of operations: boring cylinders to final diameter, surfacing deck faces, cutting threads, drilling oil passages, and honing main bearing bores. Each step requires specific tooling and setup, and each must maintain tolerances that often run to thousandths of an inch. Finally, the finished block moves to assembly, where it receives the rotating assembly, cylinder heads, and external components, transforming it into a complete engine.

Examples of improved gasket materials and torque specs show how even unglamorous details mattered. Early engine designs often struggled with oil and coolant leaks; gasket technologies simply couldn't handle the heat, pressure, and clamping forces involved. The Small Block benefited from advances in composite gasket materials that sealed more reliably and tolerated higher temperatures. Head gasket designs incorporated more sophisticated sealing rings around combustion chambers and coolant passages. Torque specifications evolved through testing, balancing sufficient clamping force against the risk of distorting the aluminum or cast-iron surfaces being fastened.

A brief comparison between a hand-built prototype engine and a typical production engine highlights where process discipline replaced individual craftsmanship. Prototype engines often received careful hand-fitting: technicians would measure clearances, adjust parts, even file or reshape components to achieve ideal fits. Production engines couldn't afford that luxury; parts had to fit correctly right out of the box. That demanded tighter manufacturing tolerances and standardization, which made craftsmanship less necessary. In many ways, that shift from art to science was the real revolution. Eliminating the need for skilled hand-fitting opened the door to mass production without sacrificing reliability.

When you build or rebuild an engine today, how much do you rely on the assumption that factory machining was generally accurate? If you're starting with a factory block, you probably assume the bores are reasonably round and straight, the deck is relatively flat, and the main bearing bores align properly. Those assumptions are usually safe, but not always. Factory tolerances, while tight by mass-production standards, still left room for variation. High-performance builds benefit from verifying critical dimensions and, where necessary, having machine work done to bring everything to optimal spec.

Experienced builders check deck flatness with a straightedge and feeler gauges. They measure bore taper and out-of-round to determine whether cylinders need boring or honing. They verify main bearing bore alignment and check that the crankshaft journals are smooth and dimensionally correct. These steps aren't mistrust of factory work; they're a recognition that even good production processes have tolerances, and pushing an engine beyond its original design parameters often requires further tightening those tolerances.

Does knowing the manufacturing story change how you view "factory stock" as a baseline for performance and durability? It should. Factory stock means an engine built to tolerances that balanced cost, reliability, and performance for typical use. That's an excellent foundation, but it doesn't mean the engine was optimized for maximum output or prepared for sustained high-rpm operation. Understanding production realities helps you calibrate your expectations and plan modifications intelligently.

Next time you inspect a Small Block, or study photos of one, take note of machine surfaces: deck finish, main caps, lifter bores. Make a short list of which features you would always re-measure and verify before a performance build, even on a supposedly "good" core. You're effectively recreating your own quality-control checklist on top of the original factory work, adding another layer of assurance before you commit time and money to assembly.

Section 4.6: From Line to Track: Early Performance and Racing Applications

The same manufacturing system that fed Chevrolet showrooms also, sometimes indirectly, supplied race tracks across America. As soon as racers realized that a mass-produced engine could deliver impressive power and stay together under stress, the Small Block began its long march into organized and grassroots competition. This journey would cement its reputation for performance for generations.

Early performance applications emerged almost immediately. Dealers and independent speed shops, always attuned to customer demand for more power, began offering tuning services, cam swaps, and dual-quad or multi-carb setups on essentially stock short blocks. These weren't full race builds requiring exotic parts or extensive machine work; they were bolt-on modifications that unlocked additional performance from factory components. The fact that such modifications worked reliably spoke to the underlying strength of the production engine.

Racers discovered that the Small Block's lightweight, compact size, and robust bottom end made it an ideal foundation for circle track racing, drag racing, and endurance events. On short-track ovals, the engine's combination of torque and clean revs gave drivers responsive power out of corners. On drag strips, quick-revving characteristics and strong mid-range punch delivered competitive elapsed times. In endurance racing, durability mattered most, and the Small Block proved it could withstand hours of hard use without grenading.

The key connection back to the assembly line is this: racers trusted the factory parts. That confidence is remarkable when you think about it. These were production components, built by the thousands on assembly lines focused on efficiency and consistency, rather than hand-selected pieces crafted by master artisans. Yet racers had enough faith in the underlying quality of the materials and manufacturing to spin a production short block to high rpm with relatively modest changes and expect it to survive. That trust spoke volumes about Chevrolet's success in translating precision engineering into volume production.

These early successes began a feedback cycle that would accelerate through the 1960s and beyond. Racing victories boosted sales, and customers wanted the engine they'd seen winning races. Higher sales justified increased production, which lowered costs and made the engine even more attractive. More production supported even broader availability of parts, both from Chevrolet and the growing aftermarket. Eventually, this ecosystem would transform the Small Block into the default choice for American performance, a status explored in later chapters.

Early stock-car or drag-strip wins, where the engine remained largely factory-spec internally, demonstrated the production engine's capabilities. NASCAR's early years, before factories fielded highly modified race-specific engines, saw competitors running engines close to showroom configuration, maybe with different carburetion, freer-flowing exhaust, and revised ignition timing, but using factory blocks, cranks, and rotating assemblies. The Small Block's success in that environment validated its design under the most demanding conditions.

The History of the Small Block Chevy Motor

Stories of regional racers pulling engines directly out of wrecked or retired street cars and repurposing them for competition, with minimal teardown, illustrate the accessibility that set the Small Block apart. You didn't need a specialty race engine built by a renowned shop; you could start with a salvage-yard 265, verify the condition of critical components, add performance parts, and go racing. That democratization of speed opened doors for enthusiasts who couldn't afford exotic equipment but had the determination to compete.

Initial factory-sanctioned or quietly supported performance programs tested the production line's ability to produce stronger components when needed. Chevrolet walked a careful line: officially, the company adhered to the Automobile Manufacturers Association's 1957 ban on factory involvement in racing. Unofficially, engineering support flowed to select racers, and certain parts,heavier-duty components, improved oiling systems, better breathing cylinder heads, found their way into the hands of competitors. These quasi-factory programs pushed manufacturing capabilities, proving that the assembly line could produce performance parts at scale when demand justified it.

In your mind, what makes a "good" engine platform for racing? Is it primarily design, or the theoretical performance potential locked in the architecture? Or does production consistency matter just as much? The Small Block's story suggests both are essential. A brilliant design that can't be built consistently won't succeed in racing, where reliability often determines results more than peak power. Conversely, a superbly manufactured engine with mediocre design eventually hits performance limits. The Small Block succeeded because it combined solid engineering with production discipline, each reinforcing the other.

If you were building a period-style race car today, how important would it be to start with a correct-era block versus any later Small Block or crate engine? That depends on your goals. If you're pursuing authenticity, a car that represents what actually competed in the 1950s or 1960s, period-correct components matter significantly. They're part of the story you're telling. If you're focused purely on performance or simply want a fun, reliable vintage racer, modern components offer advantages in parts availability, durability, and often power potential. Neither approach is wrong, but they serve different purposes and satisfy different priorities.

Choose one early form of competition, NASCAR stock cars, local dirt-track racing, or 1950s drag racing, and research how competitors described the Small Block at the time. Note at least two traits they valued most: durability, willingness to rev, easy parts access, tuning flexibility, whatever characteristics appear repeatedly in period accounts. Then compare that list with what you personally look for in a performance build today. You'll likely find that the fundamentals haven't changed much. The specific parts and technologies have evolved dramatically. Still, the underlying qualities that make an engine "good" for performance remain remarkably consistent: reliability under stress, responsive power delivery, broad parts support, and the ability to be tuned and modified without constant drama.

In this chapter, you've traced how the Small Block moved from drawing board to assembly line, then out into the world where it would prove itself in the harshest tests imaginable. You saw how the 1955 launch demanded rapid scaling and process refinement under intense market pressure, and how Chevrolet had to solve foundry, machining, training, and logistics challenges simultaneously while dealers waited and competitors watched. You recognized how early market reception and period road tests confirmed that Chevrolet's V8 was not just new, but genuinely competitive, delivering performance and drivability that won over skeptics and enthusiasts alike.

You traced how the 1957 fuel-injected 283 and its one-horsepower-per-cubic-inch claim showcased the engine's performance headroom within a production framework, proving that mass manufacturing could support genuine sophistication when engineers pushed hard enough. You saw how hitting major volume milestones, including the million-engine mark, transformed the Small Block into a foundational product for Chevrolet, no longer an experiment but a proven platform worth decades of investment.

The History of the Small Block Chevy Motor

You identified the manufacturing innovations in casting, machining, and quality control that made high-volume precision possible, understanding that the Small Block's legend rests as much on production discipline as on engineering cleverness. And finally, you saw how early performance and racing use validated the assembly line's output in the harshest real-world conditions, building trust that would fuel decades of competition success and aftermarket development.

The core theme of this chapter is simple but powerful: revolutionary design becomes historically significant only when it can be built, repeatedly, at scale. The Small Block Chevy's legend rests as much on the discipline of its assembly lines as on the ingenuity of its engineers.

Understanding that connection helps you see why these engines remain so trusted as both historical artifacts and performance foundations. When you're evaluating a vintage Small Block for restoration, or planning a build around a modern descendant, you're not just working with a good design; you're benefiting from decades of manufacturing refinement that turned that design into a reliable, repeatable reality.

Now that you've seen how Chevrolet mastered the challenge of building the Small Block in massive numbers, the next question naturally follows: What did they do with all those engines? In the following chapter, you'll follow the Small Block into the vehicles that defined mid-century America, the Bel Air, Impala, Corvette, Camaro, Nova, and more, and explore how this one engine family reshaped the character of Chevrolet's entire passenger car lineup. You'll see how the same basic powerplant could transform a family sedan into a performance icon, save America's sports car, and create the muscle car era that still captures imaginations today.

Each time you hear a Small Block fire up, you're not just listening to a clever V8 design; you're hearing the echo of thousands of workers, countless production decisions, and a manufacturing gamble that paid off beyond anyone's expectations. The assembly line didn't just build engines; it built trust. And that trust made it possible for the Small Block to become the heartbeat of American passenger cars, a role it would embrace with enthusiasm in the years that followed.

Chapter 5: The Heartbeat of America's Passenger Cars

On a quiet suburban street in 1955, the difference was obvious before you ever saw the car. The familiar hum of six-cylinder family sedans was broken by something new, a deeper, sharper exhaust note rolling down the block. When the Chevrolet Bel Air turned the corner, it looked like any other well-trimmed family car. Chrome gleamed in the afternoon sun, whitewall tires tracked straight down the pavement, and neighbors waved from their porches. But under its hood, the Small Block V8 had quietly rewritten the rules of what a "family Chevrolet" could be.

Owners who had grown up with modest, skillful powerplants, engines that did their job without complaint but never quite inspired enthusiasm, suddenly found themselves with genuine performance at the touch of the throttle. That change didn't just sell cars. It changed habits and expectations and eventually helped define an entire era of American car culture.

In the previous chapter, we saw how Chevrolet solved the enormous challenge of building the Small Block at scale, turning a bold engineering concept into a production reality on the assembly line. Now we follow that engine from the factory into the cars that made it famous. This chapter sits at the center of the book's narrative arc: it connects the Small Block's technical brilliance and manufacturing success to its real-world impact in driveways, highways, and showrooms across America. From here, the story will naturally lead into motorsports dominance and custom culture, but it all starts with how this engine transformed Chevrolet's mainstream passenger cars.

Section 5.1: The 1955 Bel Air: When the Family Car Found Its Voice

The 1955 Chevrolet Bel Air became the first widely recognized showcase for the Small Block V8, turning a handsome but otherwise conventional family car into a machine capable of genuine performance. In doing so, it set the tone for everything that followed. Ask any enthusiast to describe a "classic Chevy," and chances are good they'll describe a Tri-Five with a V8 under the hood. That instinct isn't accidental; it reflects a moment when styling, engineering, and timing converged to create something that felt both familiar and revolutionary.

Chevrolet positioned the new 265-cubic-inch V8 in the 1955 lineup alongside the tried-and-true Stovebolt six, the inline powerplant that had served the company faithfully for decades. The six remained the standard offering, particularly for budget-conscious buyers who prioritized fuel economy and low purchase price. But the V8 option, available across the range from the base 150 to the top-trim Bel Air, represented something different: a chance to own a modern, powerful automobile without stepping up to a more expensive brand.

The combination worked because Chevrolet didn't just drop a new engine into an old chassis. The 1955 model year brought fresh, clean styling that broke decisively with the rounded forms of early 1950s design. The body sat lower, the greenhouse felt more open, and the overall proportions suggested speed even when the car was standing still. Meanwhile, chassis updates, including an optional Powerglide automatic transmission and improved suspension

The History of the Small Block Chevy Motor

geometry, meant the car could actually use the power the Small Block delivered.

The numbers tell part of the story. A 1954 Chevrolet sedan with the base six-cylinder engine produced 115 horsepower and weighed roughly 3,200 pounds. Acceleration from zero to sixty took something north of eighteen seconds, adequate for merging onto highways but hardly thrilling. The 1955 Bel Air with the base 265 V8, by contrast, made 162 horsepower with a two-barrel carburetor, or 180 horsepower with the optional four-barrel setup and dual exhaust. Curb weight increased only modestly, to around 3,300 pounds. The result? Zero-to-sixty times dropped into the twelve-second range for the four-barrel car, and suddenly passing slower traffic or merging into fast-moving traffic became effortless rather than nerve-wracking.

Contemporary automotive journalists noticed immediately. Motor Trend praised the V8 Bel Air's smoothness, noting that the engine felt refined at idle and delivered power in a progressive, controllable way, not the lumpy, temperamental character some associate with high-performance engines. Popular Mechanics highlighted the passing power, observing that the Small Block gave family sedan drivers a reserve of capability that simply hadn't existed before. These weren't hot rods or sports cars; they were practical, well-trimmed automobiles that happened to be genuinely quick.

The Tri-Five Chevrolets,'55, '56, and '57, benefited from incremental improvements each year. The 1956 models received a displacement bump to 265, with power ratings climbing as high as 205 horsepower in top trim. By 1957, the 283 cubic-inch version arrived, and with it came the fuel-injected "fuelie" option that achieved the magical benchmark of one horsepower per cubic inch. Even in carbureted form, the 283 delivered between 185 and 270 horsepower, depending on configuration, giving buyers a spectrum of choices ranging from a mild-mannered daily driver to a genuine street performer.

The History of the Small Block Chevy Motor

What made these engines work so well in family car applications was the same compact, lightweight design philosophy we explored in earlier chapters. The Small Block's thin-wall casting technology and efficient packaging meant it didn't overwhelm the front suspension or demand massive cooling systems. Engineers had left enough room in the engine bay for routine maintenance, and the simple valvetrain meant mechanics across the country could service these engines without exotic tools or specialized training. The Small Block delivered strong acceleration without ruining ride quality or turning fuel consumption into a financial burden, at least by the standards of V8 performance.

Today, restorers face an interesting choice when approaching a Tri-Five project. Many of these cars have been on the road for nearly seventy years, and along the way, engines have been swapped, modified, or replaced as original powerplants wore out or owners chased more power. A significant number of '55–'57 Chevrolets now run later 350 cubic-inch Small Blocks, a swap that's mechanically straightforward and delivers more torque and horsepower than the original 265 or 283.

There's nothing wrong with that approach, especially if the goal is a reliable driver that can keep pace with modern traffic. But there's also something to be said for preserving or returning to an original-style powerplant. A correctly built 265 or 283 has a different character, slightly higher-winding, a bit more eager to rev, and lighter on its feet than the torque-heavy nature of a 350. Authenticity matters to some enthusiasts, not out of rigid dogma, but because these earlier engines genuinely feel period-correct when you're driving the car as it was originally intended.

Verifying originality requires a bit of detective work. Factory engine codes stamped on the block, casting numbers on major components, and visual details like valve cover designs and accessory mounting points all provide clues about whether you're looking at a numbers-matching engine or a later swap. Period-correct carburetion, ignition systems, and exhaust configurations complete the picture. For restorers aiming at concours-level accuracy, these details matter enormously. For builders focused on driving enjoyment, they matter less, but it's still worth understanding what you're starting with and what choices previous owners made.

When you think about a "classic Chevy," do you instinctively imagine a Tri-Five with a V8? If so, why do you think that image endures so strongly, even among people who aren't deep enthusiasts? Perhaps it's because the '55 Bel Air and its siblings hit a sweet spot, attractive enough to catch the eye, fast enough to earn respect, and common enough that almost everyone either knew someone who owned one or had ridden in one at some point. These weren't trailer queens or museum pieces. They were cars that went to work, hauled groceries, took families on vacation, and occasionally ran stoplight-to-stoplight against the neighbor's Ford down at the edge of town.

Section 5.2: Full-Size Performance, Impala, Caprice, and the Birth of the Muscle Sedan

The Small Block didn't just live under the hoods of flashy coupes; it became the backbone of Chevrolet's full-size lineup, Impala, Caprice, Biscayne, and others, creating a new category: the comfortable, well-trimmed "muscle sedan." These were big, family-oriented cars that could deliver surprising speed and capability without sacrificing practicality. In an era when most buyers still thought of performance as the exclusive domain of sports cars and hot rods, the idea that you could order a four-door sedan with genuine hustle was quietly revolutionary.

The evolution of Small Block power in full-size Chevrolets traced a steady arc from the late 1950s through the height of the muscle era. Early full-size models equipped with the 283 offered respectable performance, particularly when ordered with the optional four-barrel carburetor and performance rear axle ratios. By the early 1960s, the 327-cubic-inch version arrived, bringing a new level of refinement. The 327 delivered strong low-end torque for effortless highway cruising, but it also had enough top-end breathing to feel lively when you asked for more.

The History of the Small Block Chevy Motor

Chevrolet's strategy during this period was remarkably flexible. Within the same chassis, buyers could choose from an inline six (for maximum economy), a mild Small Block V8 (for balanced performance and reasonable fuel costs), or a Big Block V8 (for maximum straight-line thrust). This tiered approach let the company target different segments: budget-conscious families appreciated the six, performance-minded drivers who still needed four doors and a usable trunk gravitated toward the Small Block, and those who wanted bragging rights or serious towing capacity stepped up to the Big Block.

The Small Block configurations in full-size cars were surprisingly diverse. A base 283 or 327 with a two-barrel carburetor and modest compression delivered smooth, reliable power for daily driving and occasional highway trips. Step up to the four-barrel versions, and you gain noticeably sharper throttle response and better passing acceleration. Order the car with a performance axle ratio, say, 3.55:1 or 3.73:1, instead of the standard 2.73:1 economy gears, and the transformation was dramatic. The same engine that felt relaxed and understated in one configuration became genuinely quick in another, capable of running with, or occasionally embarrassing, smaller, sportier models.

Suspension, braking, and rear-axle options were often tied to engine choices, creating packages that made sense as complete systems rather than random collections of parts. A full-size Impala ordered with a 327 and performance gearing typically came with heavier-duty springs, larger brakes, and a rear end robust enough to handle the torque. These weren't formal performance packages as later years would codify them, but savvy buyers and dealers knew which boxes to check on the order form to build a car that hung together mechanically.

The interplay created some surprisingly capable machines even before the term "sleeper" became common enthusiast language. Picture a late-1960s Impala or Caprice equipped with a 327 or 350 and a 3.55 rear gear. From the outside, it looked like every other full-size Chevy sedan, with modest badging, chrome trim, and nothing to announce its potential. But roll into the throttle from a stoplight, and that big sedan moved with authority that caught people off guard. The Small Block's compact dimensions meant weight distribution stayed reasonable, handling remained predictable, and the car didn't become nose-heavy or clumsy the way some big-block intermediates could feel.

Contemporary owners appreciated these cars for reasons unrelated to street racing. Towing capability mattered: a full-size wagon with a Small Block V8 could pull a small camper or boat trailer without straining. Highway passing was safer and less stressful when you had torque on tap. Long-term durability was a given. Small Blocks ran for hundreds of thousands of miles with routine maintenance, and when they eventually wore out, rebuilding or replacing them was straightforward and affordable.

Period literature captures this practical satisfaction. Road tests from the mid-1960s frequently praised full-size Chevrolets for their balance of comfort, space, and performance. A family could load up for vacation, cruise at seventy miles per hour all day without drama, and arrive at the destination with fuel still in the tank and confidence that the car would make the return trip just as reliably. That kind of everyday capability doesn't make headlines, but it builds loyalty, and it explains why so many of these cars stayed on the road for decades.

Modern enthusiasts have discovered that full-size Chevrolets make excellent platforms for subtle performance builds. The term "sleeper" fits perfectly: outwardly stock appearance, strong but externally unremarkable Small Block under the hood, and handling upgrades that keep the car composed without shouting about it. These builds appeal to drivers who prefer understated competence over flashy image, and who enjoy the look on someone's face when a four-door sedan with hubcaps walks away from their late-model muscle car at a traffic light.

When you think about performance Chevrolets, do you automatically jump to Camaros and Chevelles, or do you also include full-size Impalas and Caprices? If you tend to overlook the big cars, it's worth asking what that says about how marketing shapes our memory. Chevrolet's advertising in the muscle era naturally focused on youth-oriented models with bold graphics and aggressive names. The full-size cars sold in huge numbers, but they were pitched to adults with families, mortgages, and practical needs. As a result, they don't occupy the same mythic space in car culture, but that doesn't mean they weren't capable, well-engineered automobiles.

If you were building a long-distance cruiser today, would you choose a Small Block over a larger displacement engine for reasons of balance, fuel economy, or ease of maintenance? There's no single right answer, but it's a question worth considering. A well-built 350 delivers ample power for highway cruising, doesn't stress modern cooling or fuel systems, and remains simple to work on decades after it was designed. For a car that's meant to cover miles comfortably rather than post numbers on a drag strip, that combination of virtues is hard to beat.

For anyone considering a full-size Chevy restoration or upgrade, start with a simple exercise: make a three-column list for the model you're interested in, say, a 1965 Impala. In the first column, note the base engine options (typically a six or a mild 283). In the second, list the optional Small Blocks (327 and 350, along with their various configurations). In the third, record the Big Block choices (396, 427, 454) if you're curious about the alternatives. Now match each combination to your priorities: cruising comfort, fuel cost, originality, or straight-line speed. The right choice depends on how and where you'll actually use the car, and being honest about that up front will save you from chasing someone else's dream instead of building your own.

Section 5.3: Saving the Sports Car, The Corvette, and Its Small Block Partnership

Without the Small Block, the Corvette's story might have ended in the mid-1950s. Instead, the engine gave Chevrolet's fiberglass sports car the performance credibility it desperately needed, turning what had been dismissed by some as a styling exercise into a legitimate performance machine. The Corvette became the Small Block's most prestigious showcase, and each new engine milestone raised the bar for American sports cars.

The Corvette's early years were troubled. Introduced in 1953 with Chevrolet's aging inline six-cylinder engine, essentially the same Stovebolt powerplant used in passenger cars, the first Corvettes produced a modest 150 horsepower and struggled to impress anyone who cared about performance. They were pretty, sure, but pretty doesn't win races or silence critics who argued that American manufacturers couldn't build a proper sports car. Sales were anemic. Production nearly ended.

Then, in 1955, everything changed. The Small Block V8 became available in the Corvette, and suddenly the car had the power to match its looks. Even the base 265-cubic-inch version with 195 horsepower transformed the driving experience, cutting zero-to-sixty times dramatically and giving the Corvette legitimate sports-car acceleration. Handling improved, too, not because of suspension changes, but because the compact, lightweight V8 didn't overburden the front end the way a heavier engine might have. Weight distribution stayed reasonable, and the car felt more balanced and eager through corners.

The real breakthrough came in 1957 with the introduction of the fuel-injected 283. Chevrolet proudly advertised "one horsepower per cubic inch", a benchmark that resonated with enthusiasts and established the Corvette as a serious performance machine, not a pretty toy for wealthy retirees. That 283-horsepower rating might seem modest by modern standards, but in context, it placed the Corvette firmly in the conversation with European sports cars costing twice as much.

Fuel injection in 1957 was exotic, temperamental, and expensive, but it delivered. Unlike carburetors, which rely on airflow to draw fuel into the intake stream, fuel injection meters precise amounts of fuel directly into each cylinder. The result is sharper throttle response, better cold starts, more consistent performance at varying altitudes and temperatures, and, crucially, more power. The "fuelie" Corvettes of the late 1950s became instant legends, sought after by collectors and enthusiasts who understood they were witnessing something genuinely special.

Of course, most Corvettes didn't get the fuel-injected engine. It was expensive, finicky to tune, and required mechanics who understood its intricacies. The carbureted Small Blocks, available in a dizzying range of configurations, served the bulk of buyers perfectly well. A 283 with a four-barrel carburetor and dual exhaust delivered 245 to 270 horsepower depending on compression and camshaft selection, and that was more than enough to make the Corvette competitive on road courses, drag strips, and mountain roads where enthusiasts gathered to test their cars and their nerve.

The high-winding 327s of the early 1960s took the partnership between Corvette and Small Block to new heights. Available from 1962 through 1965, the 327 cubic-inch engine came in multiple flavors, ranging from a mild 250-horsepower version to a solid-lifter, high-compression 375-horsepower screamer. The top-tier versions featured aggressive camshaft profiles, high-flow cylinder heads, and induction systems tuned for maximum breathing at high rpm. These engines loved to rev, and they rewarded drivers who understood how to use the full powerband, keeping the engine singing between 4,000 and 6,000 rpm, where it made its best power.

Engineers like Zora Arkus-Duntov used the Corvette as a development laboratory for high-performance Small Block technologies. Lessons learned on the track, improved oiling systems, better breathing, stronger reciprocating assemblies, filtered back into more ordinary passenger cars and the broader Chevrolet lineup. The Corvette benefited from being treated as a testbed, but so did everyone who bought a Camaro, a Chevelle, or even a full-size Impala with a well-optioned Small Block. This cross-pollination between racing development and production engineering is one reason the Small Block evolved so successfully over the decades.

The History of the Small Block Chevy Motor

The latter 350-powered Corvettes, introduced in 1968 and continuing into the 1970s, defined everyday Corvette ownership for a generation. The 350 offered a broader, flatter torque curve than the high-strung 327s, making the car more tractable in traffic and easier to live with as a daily driver. Power ranged from around 300 horsepower in base form to well over 350 horsepower in the LT-1 and other performance variants. These were engines that could idle smoothly, pull cleanly from low rpm, and still deliver thrilling acceleration when you opened them up.

A 1957 Corvette with a fuel-injected 283 is a very different machine from a 1970 Corvette with a 350 LT-1, and both are different again from a mid-1960s model with a 327. The "fuelie" cars reward patience; they start and warm up differently than carbureted versions, and throttle response has a mechanical immediacy that feels alien to anyone used to modern electronic fuel injection. Once warmed up, though, they pull with a linear, relentless urgency that's deeply satisfying. The 327-powered cars, especially the high-performance versions, are more aggressive; they want to be driven hard, and they reward high-rpm operation with a rasping exhaust note and acceleration that still feels quick today. The 350-powered cars are more refined, more flexible, and easier to live with, though perhaps less dramatic in character.

Modern Corvette owners face the same fundamental question that confronts anyone working with classic Chevrolets: preserve originality, or update for better drivability and reliability? An original Small Block Corvette, especially one with a rare engine combination, represents a piece of history. It drives, sounds, and feels different from a restomod built around a modern LS-based engine. On the other hand, an LS swap transforms the car: fuel injection means reliable starting and smooth running in all conditions, modern electronics provide better ignition timing and efficiency, and power output easily exceeds that of the original engine. What's gained is convenience and capability; what's lost is authenticity and that ineffable sense of connection to the past.

The History of the Small Block Chevy Motor

There's no universal right answer. If you think of the Corvette primarily as an "engine car", defined by its powerplant, then preserving or correctly restoring the original Small Block makes sense. If you see it as a complete package of chassis, styling, and performance, where the engine is just one component among many, then a thoughtful restomod might serve you better. The key is being honest about your priorities: originality, performance, ease of maintenance, and how you rank them.

For anyone drawn to a particular Corvette generation, C1 (1953–1962), C2 (1963–1967), or early C3 (1968–1982), take time to research the Small Block options available in a single model year. Note the differences in compression ratio, induction system, and camshaft specification. Read period road tests and owner reports to understand how these differences translate into distinct personalities behind the wheel. A 327 with 300 horsepower is not just "a little less" than one with 375 horsepower; it's a fundamentally different driving experience, tuned for different priorities and different kinds of roads.

The Corvette's success on road courses and drag strips, topics we'll explore in depth in the next chapter, owes everything to the Small Block. But even for owners who never saw a racetrack, the engine gave the car credibility, capability, and character that turned it into an American icon. Without that partnership, we might remember the Corvette as a curious footnote in automotive history, a pretty attempt that didn't quite work. Instead, it became proof that American engineers could build a world-class sports car, and the Small Block was the engine that made it possible.

Section 5.4: Camaro, Nova, Chevelle, The Small Block in the Era of Muscle and Pony Cars

As the 1960s unfolded, the Small Block became the heart of Chevrolet's answer to the pony car and muscle car movements, powering the Camaro, Nova, Chevelle, and related models through some of the most competitive years in American performance history. One basic engine architecture served drag racers, road racers, and everyday commuters across multiple nameplates and trims, proving its versatility in a way that still impresses today.

The Camaro arrived for the 1967 model year as Chevrolet's response to the Ford Mustang, which had single-handedly created the "pony car" category three years earlier. From the beginning, the Small Block fit the Camaro's mission perfectly. Base models came with inline sixes or mild Small Block V8s that delivered respectable performance and decent fuel economy. Move up the range, and you find progressively more aggressive combinations, higher compression, bigger carburetors, and more aggressive camshafts that blur the line between sporty coupe and genuine street performer.

The most famous first-generation Camaro, though, was the Z/28. Introduced in 1967 specifically to qualify the car for Sports Car Club of America (SCCA) Trans-Am road racing, the Z/28 used a 302-cubic-inch Small Block, a displacement chosen because Trans-Am rules limited engines to 305 cubic inches, and Chevrolet wanted every advantage. The 302 was built by combining the 327's 4.00-inch bore with the 283's shorter 3.00-inch stroke, creating an engine that loved high rpm and delivered its power in a narrow but potent band between 5,000 and 7,000 rpm.

On paper, the Z/28's 302 produced 290 horsepower, a figure everyone understood to be a significant understatement, chosen for insurance and racing homologation purposes. In reality, a well-tuned 302 made closer to 350 horsepower or more, with a willingness to rev that made it devastatingly effective on road courses. The engine featured solid lifters (requiring periodic valve adjustment), a high-lift camshaft, large-valve cylinder heads, and an aluminum intake manifold topped by a Holley four-barrel carburetor. It was loud, aggressive, and absolutely purposeful, an engine built to win races, not idle quietly at stoplights.

The Z/28 proved its worth immediately. Mark Donohue and Roger Penske campaigned Camaros in Trans-Am racing, winning championships and establishing the car's credentials in the crucible of professional competition. Those victories validated the entire platform and cemented the Z/28's reputation as something special, a factory-built road racer you could drive to the track, compete, and drive home.

The Chevelle and Nova told similar but distinct stories. The Chevelle, positioned as a mid-size platform, offered Small Block V8s across its lineup, from mild 307s and 327s in base models to hot 350s and eventually solid-lifter L78 and L79 396 big-blocks in SS (Super Sport) packages. But many enthusiasts preferred the Small Block Chevelles for their balance. A 350-powered Chevelle SS weighed less over the front wheels than its big-block sibling, which meant sharper turn-in, more predictable handling, and less stress on brakes and suspension components. You gave up some straight-line thrust, sure, but you gained a car that felt more cohesive and controllable, especially on twisty roads where power alone doesn't determine the fastest way through a corner.

The Nova played a different role. Smaller and lighter than the Chevelle, the Nova started life as an economy compact, but it accepted Small Block V8s with surprising ease. A Nova with a warmed-over 350, maybe with a mild performance cam, headers, a better carburetor, and a set of gears, became a classic "sleeper." Unassuming exterior, modest badging, and enough performance to surprise anyone who underestimated it. These were the cars that dominated street racing in small towns across America, where a young mechanic with basic tools and weekend time could build something genuinely quick without exotic parts or enormous expense.

The beauty of the Small Block in these platforms was its flexibility. You could build a mild street engine that ran on pump gas, idled smoothly, and got reasonable fuel economy for daily driving. Or you could go aggressive, high compression, radical cam, big carburetor, open exhaust, and create something that barely tolerated street use but absolutely dominated at the strip. Most builders landed somewhere in the middle: an engine that could be driven to work or school during the week, but that turned serious when the weekend arrived.

The History of the Small Block Chevy Motor

Insurance pressure and emissions regulations fundamentally changed the landscape in the early 1970s. Compression ratios dropped to 11:1, and higher numbers that had been common in the late 1960s fell to 9:1 or 8.5:1 as automakers responded to new unleaded fuel requirements and tightening emissions standards. Advertised horsepower ratings dropped, both because engines made less power and because the industry switched from gross to net ratings (measured with all accessories and exhaust systems in place, rather than on a bare dyno). A 1972 Camaro with a 350 might be rated at 200 net horsepower, significantly less than the 300 gross horsepower claimed just a few years earlier.

But the Small Block remained a flexible, tunable foundation. Enthusiasts discovered that with aftermarket parts,better-flowing heads, improved induction, and modern ignition systems, you could wake up these smog-era engines and extract performance that felt close to what earlier high-compression versions delivered. The basic architecture was sound; it just needed help overcoming the restrictions imposed by emissions equipment and conservative factory tuning.

When you think about "muscle cars," do you instinctively prioritize displacement and peak horsepower, or do you consider balance, weight, and how a car feels on real roads? It's a question worth asking, because the mythology around big-blocks and cubic inches sometimes overshadows the real-world advantages of a well-built Small Block in a lighter chassis. A 396 Chevelle is an icon, no question. Still, a 350 Chevelle with the right suspension and gearing might actually be more fun to drive daily, easier to maintain, and nearly as quick in most conditions that don't involve quarter-mile timing lights.

If you were selecting a powerplant for a 1960s or early 1970s Chevy project today, would you automatically chase a big-block badge, or might a well-built Small Block better match how and where you actually drive? For most people, the Small Block is the smarter choice. Parts are more plentiful, machine shops have more experience, and the engines are simply easier to tune and live with.

They weigh less, fit better in crowded engine bays, and don't punish you at the gas pump the way a big-block inevitably will.

Pick one classic Chevy platform, Camaro, Chevelle, or Nova, and identify three factory Small Block options that were offered across a few model years. Note their rated horsepower and compression ratios, and consider how fuel availability and modern driving expectations might influence which one you'd want to run today. A 1969 Z/28 with an 11:1-compression 302 is an extraordinary machine, but it requires racing fuel or aggressive timing tuning to run on modern pump gas. A 1972 Camaro with a 9:1 350 won't impress anyone with its factory specs, but it's a blank canvas that responds beautifully to thoughtful upgrades, and it'll run all day on 91 octane without complaint.

The Small Block's success in Camaros, Chevelles, and Novas during the muscle car era wasn't just about horsepower. It was about accessibility; these were engines that regular people could afford, understand, and modify. They were forgiving enough that a novice builder could learn on them without immediately destroying expensive parts, and potent enough that an experienced tuner could extract truly impressive power. That combination made them the foundation of street performance culture in a way that no other engine family quite matched.

Section 5:5: Compact Platforms, Station Wagons, and Everyday Versatility

The Small Block didn't only live in headline-grabbing performance models. Its real claim to "heartbeat of America" status comes from how often it appeared in more modest roles, compact cars, family wagons, and entry-level models that quietly benefited from V8 power. This widespread use demonstrates the engine's adaptability in a way that's easy to overlook when the conversation always turns to Corvettes and Z/28s.

The History of the Small Block Chevy Motor

Chevrolet's compact and intermediate platforms, especially the Nova and later smaller Chevys, used the Small Block to bridge the gap between economy and capability, often as an affordable upgrade over four- and six-cylinder options. A base Nova came with a four-cylinder or six-cylinder engine that delivered adequate performance for someone who just needed reliable transportation. But step up to even a mild Small Block, say, a two-barrel 307 or 327, and the car transformed. Merging onto highways became effortless. Passing slower traffic no longer required elaborate planning. And the driving experience simply felt more confident and composed.

Station wagons represent an even better example of the Small Block's everyday versatility. These were family haulers, built to carry kids, luggage, camping gear, groceries, and whatever else needed to be moved from point A to point B. Most buyers ordered wagons with comfort and convenience in mind, air conditioning, automatic transmission, power steering, but many also checked the box for a V8 engine, and for good reason.

Towing capability mattered. A wagon equipped with an inline six could pull a small utility trailer in a pinch, but ask it to tow a camper or a boat, and the experience became stressful, with overheating, sluggish acceleration, and constant downshifting on hills. A wagon with even a modest Small Block handled those tasks with composure. The extra torque meant the engine wasn't laboring constantly, temperatures stayed under control, and the driver could focus on the road instead of worrying about mechanical distress.

Highway cruising became more pleasant, too. A six-cylinder wagon doing seventy miles per hour on an interstate was working hard, running at high rpm, burning through gas, and generating noise and vibration that wore on passengers during long trips. A Small Block–powered wagon loafed along at the same speed, the engine turning lazily below 3,000 rpm, barely breaking a sweat. Fuel economy wasn't dramatically worse; the V8's additional torque often meant it could use taller gearing, which offset some of the displacement disadvantage, and the overall refinement was noticeably better.

The History of the Small Block Chevy Motor

Long-term durability was another advantage. Small Blocks in everyday use, moderate compression, conservative tuning, regular maintenance, routinely accumulated hundreds of thousands of miles. When they eventually wore out, rebuilding or replacing them was straightforward and affordable. Parts were available at any auto parts store, mechanics across the country understood them, and the rebuild process was well documented in shop manuals and enthusiast literature. This wasn't exotic machinery requiring specialists and expensive tooling; it was a robust, well-understood engine that regular shops could service competently.

Consider a mid-1960s or 1970s Chevy wagon ordered with a Small Block V8 for towing a small camper or boat. The owner valued torque and reliability more than bragging rights or stoplight performance. Maybe it was a salesman who logged thousands of highway miles every month. Maybe it was a family headed to the lake every summer, boat trailer in tow. These engines didn't get written up in car magazines, and they didn't win races, but they did their jobs year after year, mile after mile, without drama.

Or look at a base-model Nova where a Small Block was a mid-level or top-level option. The car might have been ordered by someone who appreciated a bit of extra power but didn't need or want the full SS treatment with its flashier badging and stiffer suspension. That choice fundamentally changed the ownership experience: the car felt more capable, more confident, and more enjoyable to drive, without announcing itself or demanding constant attention.

Modern families have discovered that restored Small Block–powered wagons make remarkably capable road-trip vehicles. They're spacious, comfortable, and distinctively styled, turning heads and starting conversations. With modern radial tires, updated suspension bushings, and an overdrive transmission (either a period-correct option or a thoughtful retrofit), these cars can comfortably and reliably cover long distances. The experience isn't the same as driving a modern SUV; you're more aware of the road, more engaged with the act of driving, but for many enthusiasts, that's exactly the point.

The History of the Small Block Chevy Motor

How often do you notice interesting engine choices in everyday-looking cars at shows, wagons, four-doors, and base models? Do you tend to walk past them on the way to the flashier coupes? If so, you're missing stories that are often more interesting than the twentieth identical SS clone on display. The wagon that's been in one family for forty years, used for everything from daily commutes to cross-country vacations. The four-door sedan that an enthusiast's grandfather ordered new, specced exactly the way he wanted it, and maintained meticulously for decades. These cars represent real history, not fantasy builds assembled from catalogs.

If you're building a car primarily for real-world use, errands, family trips, and occasional towing, what qualities matter most in an engine: peak horsepower, low-end torque, ease of parts sourcing, or fuel economy? For most people in this situation, the answer tilts heavily toward torque, reliability, and simplicity. A 350 with a mild cam, good heads, and a well-matched induction system delivers ample power for highway passing and occasional towing, runs on pump gas without complaint, and remains easy to maintain. That's not romantic or flashy, but it's smart, and it creates a car you'll actually want to drive regularly rather than one that sits in the garage because it's too aggressive or temperamental for ordinary use.

If you attend shows or cruise nights, deliberately seek out one "ordinary" Small Block–powered Chevy and spend time studying its engine bay, options, and owner's story. Notice how often practicality and personal history shape these builds. The owner isn't trying to recreate a magazine feature or win a trophy; they're preserving something that matters to them personally, and the car reflects that authenticity. These are often the most satisfying conversations at any event, because they're about real experiences with real cars, not aspirational fantasies about what might have been.

Section 5.6: From Main Street to Midnight: Cultural Impact on Cruising and Street Life

As the Small Block spread through Chevrolet's passenger car lineup, it didn't just change spec sheets; it changed how people used their cars. Main Street cruising, impromptu street races, and late-night gatherings in parking lots all developed a distinct soundtrack, and it was usually a V8 Chevy idle. Small Block–powered Chevrolets helped define American car culture in the 1950s, 1960s, and beyond, well before organized racing or formal car clubs entered the scene.

The transformation occurred because affordable V8 power fundamentally changed youth culture. Suddenly, speed and personalization were accessible to a much wider slice of the population. You didn't need to be wealthy or mechanically gifted to own a car with genuine performance potential. A used Chevy with a Small Block cost less than a new economy car; parts were cheap and plentiful, and the basic operations, tune-ups, carburetor adjustments, and gear swaps could be learned from friends, magazines, or trial and error in a driveway. This democratization turned otherwise ordinary passenger cars into platforms for identity and social life.

Specific models became synonymous with specific activities and scenes. The Tri-Five Bel Airs, with their clean styling and available V8 power, were among the first to establish the template: modified but not radical, quick but not exotic, distinctive but not unapproachable. Early Impalas followed a similar path: big, comfortable coupes and convertibles that looked sharp, sounded good, and moved with authority when you wanted them to.

By the mid-1960s, Novas, Chevelles, and Camaros joined the scene. Novas became the quintessential sleeper: plain wrappers hiding potent engines, driven by kids who learned to wrench in high school shop classes and spent weekends hunting for speed parts at swap meets. Chevelles split the difference between muscle and manners, quick enough to be taken seriously, refined enough to be driven daily. Camaros, especially Z/28s and SS models, represented youth-oriented performance in its purest form: loud, fast, and unapologetically aggressive.

Cruising strips emerged in towns and cities across America. The routine was simple: drive a loop that everyone knew, stopping at certain parking lots or drive-ins where the scene congregated. You'd see who showed up, what they were driving, and what modifications had been made since last weekend. Conversations happened through car windows, bench racing about whose car was faster, debates about carburetor jetting, stories about close calls with the cops or near-misses with telephone poles. The cars were social currency; what you drove and how you drove it said something about who you were.

Informal drag racing happened naturally. Two cars pull up to a red light. Drivers make eye contact. Engines get revved, sometimes subtly, sometimes blatantly. The light turns green, and for the next few hundred feet, nothing else exists except throttle position, shift points, and traction. These weren't organized events with timers and safety equipment; they were spontaneous tests of nerve and machinery that blurred the line between transportation and recreation.

The ubiquity of the Small Block made it "the default engine" in many of these scenes. You could find parts anywhere, junkyards, speed shops, the guy down the street who always seemed to have a garage full of spare blocks and heads. Knowledge was shared freely: someone figured out that boring a 327 to 4.030 inches and running a high-lift cam woke the engine up dramatically, and within weeks, that information had spread through the local network. The mechanical simplicity meant you could actually work on these engines with basic tools, without computer scanners or specialized training.

Imagine a Friday night on a typical American cruise strip in the mid-1960s. Small Block–powered Chevrolets line up at stoplights: a primer-gray '57 Bel Air with twice-pipes and wide rear tires, a Butternut Yellow '66 Nova that looks stock until the hood opens, a '68 Camaro SS with headers loud enough to set off car alarms three blocks away. The rivalries are casual but real; nobody wants to be the one who gets walked on Main Street in front of everyone. Friends and onlookers learn to distinguish cars by sound and stance: that lumpy idle means a serious cam, those raked rear springs mean drag strip aspirations, that straight-pipe exhaust means someone who either doesn't care about attention or actively wants it.

The History of the Small Block Chevy Motor

Local enthusiast scenes are often built around a single model. In a small town, you might find a cluster of street-driven Camaros or Novas, their owners connected by a shared enthusiasm for a particular platform. The ease of swapping and tuning Small Blocks kept these cars running and relevant for decades. When an engine wore out, you didn't park the car; you found another 350 from a wrecked wagon or a worn-out truck, rebuilt it in someone's garage over a few weekends, and dropped it back in. The cars stayed on the street because maintaining them was economically and mechanically feasible.

Some of these street drivers eventually moved into formal drag racing or circle track events, a natural progression from stoplight contests to organized competition. The same engine combinations that worked on the street formed the foundation for entry-level race classes. A budget-minded racer could pull a 350 out of a passenger car, add better rods and pistons, port the heads, tune the carburetor, and have a viable engine for Stock or Super Stock racing. The knowledge and skills developed on the street translated directly to the track, and vice versa. This cross-pollination strengthened both worlds and helped build the small-block Chevy's reputation as the engine that could do anything.

Think back to the first time you really noticed a car because of the way it sounded or accelerated. Was it a particular model, or simply "a V8 Chevy" that stuck in your mind? For many people, the specifics blur: they remember the exhaust note, the sense of barely controlled power, the image of a car that looked fast even standing still. That emotional imprint, formed decades ago, often drives decisions today about what to buy, build, or restore.

In your own region or generation, what played the role that Small Block–powered Chevrolets did in the 1950s and 1960s, an accessible, modifiable car that young drivers could claim as their own? The answer varies, Fox-body Mustangs in some areas, older Japanese imports in others, front-wheel-drive sport compacts in still others, but the underlying pattern remains consistent: young people gravitate toward affordable vehicles they can personalize, modify, and use as expressions of identity. The Small Block Chevy established that template, which still resonates.

If you're building or owning a Small Block–powered passenger car today, consider one way you might intentionally preserve or recreate a piece of that cultural experience, whether it's the exhaust note, the stance, or simply taking the long way home on a summer evening. Maybe it's leaving the stereo off and just listening to the engine. Maybe it's finding a local cruise night and actually participating instead of just observing. Maybe it's taking a younger enthusiast for a ride and explaining why this car, this engine, matters beyond horsepower numbers or auction prices.

The Small Block's cultural impact extends far beyond mechanical specifications. It soundtracked a generation's coming of age. It taught millions of people basic automotive skills. It enabled social interactions and created communities that persist to this day. When we talk about the Small Block as "the heartbeat of America," we're not just referencing its commercial success or engineering excellence; we're acknowledging its role in shaping how people lived, what they valued, and how they expressed themselves.

In this chapter, we followed the Small Block Chevy out of the engineering lab and into the real world of Chevrolet's passenger cars. We saw how it transformed the 1955 Bel Air from a stylish family car into a performance icon, gave full-size Impalas and Caprices unexpected muscle, and quite literally saved the Corvette by supplying the power it always needed. We traced its influence through the Camaro, Nova, and Chevelle during the height of the pony and muscle car eras, then recognized its quieter but equally important role in compacts, wagons, and everyday family vehicles. Finally, we looked at how these Small Block–powered Chevrolets shaped American cruising, informal street racing, and the broader culture of how people used and enjoyed their cars.

The Small Block Chevy became "the heartbeat of America" not just because it was powerful or well-engineered, but because it was everywhere, from glamorous Corvettes to workhorse wagons. Its adaptability allowed Chevrolet to offer a common mechanical core across wildly different body styles and buyer segments, turning one engine family into a unifying thread through decades of passenger car history. Understanding where and how the Small Block appeared in these cars gives today's enthusiasts practical tools for accurate restoration, informed modification, and a deeper appreciation of why certain combinations feel "right" when you encounter them, whether in a show field or your own garage.

Up to this point, we've focused on the street: how the Small Block reshaped family cars, sports cars, and everyday driving. But while late-night cruises and stoplight duels were important proving grounds, they were only part of the story. In the next chapter, we'll follow the Small Block onto the formal stages of competition, NASCAR ovals, drag strips, road courses, and off-road events, where the same qualities that made it a brilliant passenger car engine would underpin a racing dynasty stretching across seven decades.

By the mid-1960s, you could stand on almost any American street corner and hear it: the rolling burble of a Small Block Chevy heading somewhere, a date, a race, a road trip, or just a run to the store. That sound, coming from Bel Airs, Impalas, Corvettes, Camaros, Novas, and Chevelles, announced more than an engine. It signaled a new idea of what an ordinary car could be. Those same engines would soon be pushed harder, tuned sharper, and tested under the harsh lights of competition. The streets had proven the Small Block's appeal; now the tracks would prove just how far its potential could be stretched.

Chapter 6: Track Domination, A Racing Dynasty

A Saturday night short track in the late 1960s with dust hanging in the air, grandstands packed, and a field of stock cars idling on the grid. When the starter's flag drops, a familiar sound rips through the noise, a hard-edged, urgent V8 bark that, from local bullrings to Daytona and Le Mans, became instantly recognizable. That sound, more often than not, came from a Small Block Chevy.

In the previous chapter, we saw how the Small Block reshaped Chevrolet's passenger car lineup and helped define American performance on the street. Now we move into the proving ground that truly separates legend from hype: competition. On the track, weaknesses are exposed quickly, and strengths are amplified. The Small Block's racing story explains why its basic architecture survived, adapted, and dominated for decades.

By the end of this chapter, you'll be able to trace the Small Block's rise in NASCAR, drag racing, road racing, and off-road competition. You'll understand how racing demands drove key components and design improvements. You'll recognize how competition engines influenced the aftermarket ecosystem you can buy from today. And you'll be able to identify race-bred modifications that translate effectively to street and track-day builds, and which ones don't.

Section 6.1: NASCAR Supremacy, From Short Tracks to the Super Speedway

The Small Block Chevy became a cornerstone of stock car racing almost from the moment it arrived. When Chevrolet introduced the 265 cubic-inch V8 in 1955, NASCAR was still young, a scrappy, regional series where "strictly stock" meant something closer to its literal definition than it does today. The Small Block's compact dimensions, strong bottom end, and efficient breathing characteristics made it ideal for sustained high-RPM oval racing, and racers noticed immediately.

Those early wins weren't accidents. The engine's compact size allowed better weight distribution in the chassis. Its overhead valve design breathed more efficiently than the flathead and early OHV engines it competed against. And perhaps most importantly, its cast-iron block with deep skirts and four-bolt main bearing caps (on performance variants) provided the structural integrity needed to survive hours of racing at high RPM and sustained loading through the turns.

The interplay between factory engineering and independent race shops defined the Small Block's evolution in NASCAR. Chevrolet's official involvement in racing waxed and waned over the decades, sometimes providing direct support, other times stepping back due to corporate policy or the Automobile Manufacturers Association's 1957 racing "ban". Yet, regardless of factory participation, the basic Small Block architecture remained a favorite among builders. Why? Because it responded predictably to modifications, parts were becoming increasingly available, and the fundamental design was sound enough to handle serious abuse.

Consider a representative Cup-caliber Small Block from the heyday of pushrod NASCAR racing: high-compression pistons squeezing out every bit of efficiency from premium fuel, an aggressive

The History of the Small Block Chevy Motor

solid roller camshaft with duration and lift figures that would make a street engine nearly undriveable, hand-ported cylinder heads with intake and exhaust ports meticulously shaped for maximum flow, and rotating assemblies balanced so precisely that the engine could survive 500 miles at race RPM without self-destructing. These weren't just tweaked production engines; they were purpose-built racing machines that happened to share the Small Block's basic architecture.

NASCAR's governing bodies responded to the Small Block's performance potential with an evolving set of rules: restrictor plates to limit airflow and power on superspeedways, displacement limits to control cubic inch wars, and homologation requirements that tied race engines to production reality. Many of these rules were aimed, at least in part, at controlling what engines like the Small Block could do when talented builders got their hands on them.

The Small Block's NASCAR legacy extended far beyond the Cup Series. In late-model short track racing across America, Small Block Chevrolets became the default choice. Their combination of affordability, reliability, and performance potential meant a racer with modest resources could build a competitive engine, go racing on Saturday night, and, if something broke, source replacement parts without mortgaging the farm. This democratization of performance is part of what made the Small Block legendary, not just in professional circles but in grassroots racing communities where most of the real work gets done.

When you think about engines built for sustained high RPM, what characteristics matter most? Certainly, the block, main caps, and rotating assembly must withstand tremendous forces. Breathing is equally critical; at race RPM, airflow limitations translate directly into lost horsepower. Cooling becomes crucial when an engine runs wide-open for extended periods. And the Small Block's basic design addressed all of these requirements well enough that it remained competitive for decades, even as technology and competition advanced.

Take a few minutes to list three features that make an engine suitable for oval-track racing: valvetrain stability at high RPM, oil control in long, sustained corners, and cooling capacity under continuous load. Next to each feature, jot down which Small Block components or design traits address that requirement: the deep-skirt block design for rigidity, priority main oiling on later variants to ensure crankshaft lubrication, aftermarket circle track oil pans with baffles and trap doors to prevent oil starvation during cornering. This exercise helps connect abstract racing requirements to concrete engineering solutions, solutions that inform what you might build today.

Section 6.2: Drag Racing Revolution, Quarter-Mile Power for Everyone

While NASCAR showcased the Small Block's endurance, drag racing highlighted its explosive potential. The Small Block Chevy played a central role in drag racing's growth from a fringe activity to organized motorsport, from grassroots bracket racing to professional Pro Stock, by offering accessible, scalable power that thousands of racers could afford and tune.

The Small Block was uniquely well-suited to quarter-mile duty. It would rev willingly when built correctly, responding eagerly to cylinder head and camshaft upgrades. It tolerated forced induction, both superchargers and turbochargers, reasonably well, and nitrous oxide systems became a popular addition for racers chasing peak numbers. High-compression naturally aspirated builds could make serious power with the right combination of parts, and the engine's compact size allowed it to fit in everything from full-size sedans to lightweight gassers.

Drag racing's focus on peak horsepower, quick bursts of torque, and repeatable launches pushed development in specific directions. Valvetrain technology advanced rapidly as builders sought higher RPM limits and better control at those speeds. Intake manifold design evolved from mild dual-plane castings to high-rise single-plane race intakes and even radical tunnel-ram setups that placed carburetors high above the engine. Ignition systems moved from points to electronic and eventually to sophisticated programmable units that could deliver precisely timed spark under extreme conditions.

Period-correct 1960s gasser builds illustrate the era's approach: a high-rise single-plane intake manifold, a big solid-lifter camshaft with aggressive timing, tunnel-ram experiments that tested the limits of intake tract length and plenum volume, and steep rear gears that sacrificed streetability for launch performance. These were purpose-built race engines that lived for the quarter-mile and nothing else.

Modern bracket-racer combinations tell a different story, one of refinement and consistency. A simple 350 or 383 cubic inch build making strong, repeatable power with a relatively mild hydraulic roller camshaft, good aftermarket cylinder heads with decent port work and valvetrain geometry, and a dialed-in fuel system that delivers consistent air-fuel ratios run after run. These engines might not make the peak numbers of all-out Pro Stock motors, but they win races by being predictable, reliable, and easy to tune.

Legendary drag racers built their reputations on Small Block power. Names like "Grumpy" Jenkins became synonymous with Chevrolet performance, developing innovative combinations that dominated classes and pushed rule limits. Countless others, racers whose names never made national headlines, earned local reputations and track championships with home-built Small Blocks that delivered weekend after weekend.

Bracket racing and sportsman classes democratized drag racing in a way that mirrored the Small Block's broader impact on performance culture. A working-class enthusiast could buy a used engine block, add catalog parts, and run competitive times against racers with far more resources. The playing field wasn't perfectly level; money always helps, but the Small Block's accessibility meant talent and tuning could overcome budget limitations more often than in other forms of racing.

When you imagine building a drag-focused Small Block, what pulls at you, maximum peak horsepower, or consistency and reliability? The answer reveals something about your priorities and probably your budget. All-out power requires constant maintenance, premium components, and a willingness to accept failures. Consistency requires discipline, careful tuning, and sometimes leaving power on the table to ensure the engine survives round after round. Neither approach is wrong; they're different philosophies that suit different goals.

The availability of crate engines and off-the-shelf rotating assemblies has fundamentally changed what "race motor" means to the average enthusiast. You can order a complete engine, balanced, dyno-tested, ready to drop in, and go racing without ever hearing a micrometer or degree wheel. This accessibility is the Small Block's legacy in its purest form: performance within reach.

Sketch out a hypothetical drag-oriented Small Block combo: displacement (383 stroked from a 350 block?), compression ratio (10.5:1 for pump gas or 12:1 for race fuel?), cam type (solid roller for maximum performance or hydraulic roller for durability?), induction (traditional carburetor or modern EFI?), and intended RPM range (6,500 or 7,500 rpm?). Then circle the components where racing experience clearly influenced your choices. That single-plane intake? Decades of drag racing data. Those stud girdles? Failures at high RPM taught that lesson. The rev kit in your valvetrain? Racers figured out the weak link and fixed it. Your hypothetical build stands on the shoulders of countless quarter-mile runs.

Section 6.3: Road Racing Excellence, From Trans-Am to Le Mans

In road racing, the Small Block proved that American pushrod V8s could do more than charge in a straight line; they could endure hours of high-speed cornering, braking, and acceleration while competing on the world stage. This was the arena where European skepticism about American engine design met reality, and the Small Block held its own.

The History of the Small Block Chevy Motor

The Trans-Am series of the late 1960s and early 1970s showcased Small Block power in its most refined form. Chevrolet Camaros equipped with high-winding 302-cubic-inch engines, built to meet displacement limits, competed against Ford Mustangs and other rivals in intense road-course battles. These engines weren't about brute force; they were about throttle response, a wide power band, and the ability to accelerate hard out of slow corners then maintain power through fast sweepers.

The Corvette's long campaign in sports car racing, including endurance events at Le Mans and Daytona, demonstrated the Small Block's capacity for sustained performance. Racing an engine for 24 hours straight places demands that differ fundamentally from drag racing or even oval-track competition. Thermal management becomes critical when there's no cool-down period between runs. Oil control under lateral G-loads requires sophisticated systems to ensure the pickup stays submerged regardless of cornering forces. Durability matters more than peak power when the engine must survive through the night and into the next day.

These demands drove specific engineering solutions. Dry sump oiling systems replaced conventional wet sumps, using an external reservoir and multiple scavenge pumps to maintain consistent oil supply under any condition. Improved cooling passages and more efficient water pumps addressed thermal loads. Bottom ends received upgrades: stronger connecting rods, better pistons, improved main caps and fasteners. Eventually, the LT and LS generations incorporated many of these lessons from the factory, refining the Small Block formula for modern endurance racing while maintaining the core architecture.

Consider the differences between a drag racing build and a road-race-focused build of similar displacement. The drag motor might run an aggressive single-plane intake optimized for peak power at high RPM, while the road racer uses a dual-plane or tuned-runner intake that maintains power across a broader range. Camshaft profiles differ: the drag cam prioritizes top-end power, while the road race cam balances mid-range response with high-RPM breathing. Oiling strategies diverge completely: the drag car might get by with a basic baffled pan, while the road racer needs comprehensive oil control with accusump backup systems and multiple trap doors. Cooling is barely a concern for a 10-second pass but becomes paramount for a two-hour stint.

Which is more impressive, a single, record-setting quarter-mile blast, or an engine that survives 24 hours of racing at high speed? The question isn't entirely fair; they're different achievements requiring different engineering priorities. Yet your answer probably reveals how you value performance. Do you prioritize peak achievement or sustained excellence? Both matter, but they lead to different design choices.

If you were building a dual-purpose street/track car, which road-race-derived features would you prioritize? A baffled oil pan with trap doors ensures the oil pickup remains covered during hard cornering, which is relevant even on spirited canyon drives. Improved cooling capacity helps when running the car hard on warm days or in traffic after a track session. A broader power band makes the car more enjoyable on the street while providing flexibility on track. These aren't just racing features; they're improvements that make the engine more capable in any demanding situation.

Review a spec sheet for a historic or modern road-racing Small Block, whether a factory C5R engine or an aftermarket road race build. Note features aimed at endurance rather than peak power: high-capacity oil system with dry sump configuration, upgraded cooling with additional capacity and improved flow, durable bottom end with forged components and upgraded fasteners. Next, identify which you could incorporate into a street or track-day build. The fully dry sump system might be excessive and expensive, but an improved oil pan is feasible. An auxiliary oil cooler is straightforward. An upgraded radiator is a bolt-on improvement. Racing teaches lessons; wisdom is knowing which lessons apply to your project.

Section 6.4: Off-Road and Rally, Durability in the Harshest Environments

Beyond paved circuits, the Small Block proved its toughness in off-road racing, desert competition, and rally-style events where heat, dust, vibration, and unpredictable loading punished every component. This is where the engine's fundamental robustness, the deep-skirt block design, the simple and serviceable architecture, and the cast-iron strength mattered most.

Off-road competition highlighted aspects of the Small Block's character that don't show up on dyno sheets. Block casting quality and strength became critical when the entire engine experienced shock loads from jumps and washboard surfaces. Serviceability mattered when repairs needed to happen in remote locations with limited tools. The ability to deliver usable torque at varying RPM ranges proved more valuable than peak horsepower when navigating technical terrain at unpredictable speeds.

Harsh environments introduced stresses that circuit racing didn't: air filtration became crucial when operating in dust for hours, requiring sealed induction systems and high-capacity filters. Cooling-challenged engineers, low-speed operation in high ambient temperatures meant less airflow through the radiator when thermal loads were highest. Shock loading from terrain features subjected engine mounts, accessory brackets, and internal components to forces they'd never see on smooth pavement. Moisture and debris threatened ignition systems, requiring better sealing and protection.

Specific adaptations addressed these challenges. Heavy-duty engine mounts absorbed vibration and shock loads without failing. Special oil pans with reinforced construction and well-designed pickup placement ensured oil supply even when the vehicle was at steep angles or experiencing violent motion. Tuned intake routing protected the induction system from water crossings and dust. Ignition components received additional sealing and shielding against moisture and debris.

The History of the Small Block Chevy Motor

Small Block-powered trucks and buggies competed in events like the Baja 1000 and other long-distance desert races where reliability mattered more than outright speed. Grassroots off-road builds, older Chevrolet trucks, 4x4s with swapped drivetrains, purpose-built rock crawlers, used junkyard Small Blocks or affordable crate engines for dependable torque and the confidence that parts could be sourced or repairs made even in the field.

Some of these off-road lessons fed into heavy-duty and marine versions of the Small Block family. Improved sealing techniques, more robust cooling systems, and components designed to tolerate sustained abuse found applications beyond racing. The marine Small Blocks, in particular, benefited from developments in corrosion resistance and closed-cooling systems first proven in harsh racing environments.

When you think of "reliability," what comes to mind: highway mileage accumulating predictably, or survival in extreme conditions? Both are valid measures, but they test different aspects of design. The Small Block's off-road record demonstrates that the engine could handle abuse beyond what most street applications would ever see. That's reassuring whether you plan to race in the desert or simply want confidence that your engine has capability in reserve.

If you regularly drive in hot climates, tow heavy loads, or use dirt roads, certain off-road-derived upgrades make practical sense. An upgraded radiator with additional capacity helps when ambient temperatures climb or you're working the engine hard. An auxiliary transmission cooler protects the transmission when towing. A higher-capacity oil pan with baffles provides additional oil volume and better control. Improved filtration protects against the fine dust that inevitably works its way into vehicles driven on unpaved roads. These aren't exotic racing parts; they're sensible improvements informed by competition experience.

Identify one weak link in a typical street Small Block when exposed to off-road or heavy-duty conditions. Perhaps it's marginal oiling during steep climbs, where oil sloshes away from the pickup. Or inadequate cooling capacity when towing in hot weather at sustained highway speeds. Research common fixes: a baffled oil pan with increased capacity addresses oiling concerns; an upgraded radiator with auxiliary electric fan improves cooling; a quality air filter with better sealing protects against dust ingestion. Note which would be easiest to implement on a street-driven vehicle. Usually the simplest improvements, better filtration, improved cooling capacity, deliver meaningful benefits without requiring extensive modification.

Section 6.5: Racing as R&D: How Competition Built the Aftermarket

Racing didn't just showcase the Small Block; it became an informal research and development lab that drove the creation of the modern performance aftermarket. The parts catalog you browse today, whether online or in printed form, exists because racers identified problems, developed solutions, and proved those solutions worked under the most demanding conditions possible.

Competition needs drove innovation in predictable ways. Racers needed more airflow, so cylinder head development accelerated. They needed greater strength, so forged rotating assemblies and upgraded fasteners became available. They needed better control, so improved valvetrain components, more precise camshaft grinds, and sophisticated ignition systems emerged. Each racing discipline contributed its own lessons, and small speed shops that solved these problems often grew into major brands.

Consider the evolution of aftermarket Small Block cylinder heads. Early performance heads were production castings with modest port work and slightly larger valves, improvements, but incremental ones. As racing advanced and flow bench technology became more accessible, head design grew more sophisticated. Porting techniques improved, informed by countless hours of testing and competition data. Eventually, fully CNC-ported aluminum heads with redesigned port configurations, improved valve angles, and optimized combustion chambers became available, designs that would have been unthinkable without decades of racing experience providing the knowledge base.

Camshaft development followed a similar arc. Early performance cams were relatively mild, limited by manufacturing capabilities and incomplete understanding of valve timing's effects. Circle-track racing generated data about duration and overlap's impact on power bands. Drag racing pushed lift and duration to extremes while exposing valvetrain stability issues that needed solving. Road racing emphasized broader power characteristics and reliability. Over time, this accumulated knowledge led to extensive camshaft catalogs with profiles tailored to specific applications, street/strip compromises, dedicated race grinds, economy-focused cruiser cams, and everything in between.

Many well-known aftermarket brands started as race shops, small operations run by talented mechanics and engineers who recognized opportunities to improve Small Block performance. They'd develop a better intake manifold or stronger connecting rod to solve a problem they encountered racing, then discover that other racers wanted the same solution. As demand grew, these shops evolved into manufacturers serving not just racing but the broader enthusiast market. Their racing pedigree provided credibility, and their parts carried the implicit promise that if they worked under race conditions, they'd certainly handle street duty.

Yet this relationship between racing and the aftermarket created a double-edged sword. While race-derived parts often delivered substantial performance gains, not every "race" component made sense or even functioned well in a street environment. A cam profile perfect for a circle-track engine turning 7,000 RPM might make a street car nearly undriveable, with rough idle, poor vacuum, and a power band that starts well above typical street RPM. A single-plane intake optimized for top-end power might sacrifice low and mid-range response that makes a car enjoyable in normal driving. Lightweight racing components might lack the durability for sustained street use.

When you see a part marketed as "race proven," what does that mean to you? Does it inspire confidence and provide proof that the component can handle serious abuse? Or does it raise caution and recognition that race priorities differ from street needs? Both reactions are reasonable. The wisdom lies in understanding what "race proven" actually proves for your specific application.

Think about your own goals. Are you trying to mimic a race engine, accepting the compromises that entails? Or are you borrowing just enough race technology to make your build more responsive, durable, or efficient while maintaining streetability? Neither approach is wrong, but they lead to very different parts selections and build strategies.

Section 6.6: Legends Behind the Wheel, People Who Made the Small Block Famous

Engines don't win races on their own. Drivers, builders, and teams turned the Small Block's potential into results, and their stories help explain why this engine inspires such loyalty across generations. The hardware mattered, certainly, but the human element transformed capability into achievement.

Professional racing provided high-profile examples. Drivers like Junior Johnson, Richard Petty, and Dale Earnhardt built NASCAR legacies that included countless laps behind Small Block power. Drag racing legends like "Grumpy" Jenkins and Warren Johnson became synonymous with Chevrolet performance, developing innovative combinations that dominated their classes and pushed rule makers to respond. Road racers like Dick Guldstrand helped prove that Small Block Corvettes could compete internationally against sophisticated European machinery.

Yet the Small Block's story extends far beyond professional motorsports. Local heroes, racers whose names never appeared in national magazines, earned reputations at Saturday-night short tracks across America. These builders and drivers worked regular jobs during the week, then spent evenings and weekends preparing home-

The History of the Small Block Chevy Motor

built Small Block engines that competed against rivals with bigger budgets and sometimes more sophisticated equipment. Their ingenuity and persistence wrung remarkable performance from ordinary production blocks, often matching or beating more exotic machinery through careful tuning, smart driving, and thorough understanding of their equipment.

The culture of grassroots racing created communities of shared knowledge. In pit lanes and garages, racers exchanged information about what worked and what didn't. A builder who discovered that a specific combination of cylinder heads, cam timing, and carburetor jetting delivered exceptional mid-range power would share that information, not because they expected anything in return, but because that's how the community operated. This open exchange accelerated Small Block development in ways that benefited everyone, from weekend racers to professional teams.

The relationship between engine builders and drivers deserves particular attention. A talented driver provides detailed feedback about engine behavior: how it responds to throttle inputs, where power comes on, whether it pulls consistently across the RPM range, and how it behaves when hot versus cold. A skilled builder translates this subjective feedback into objective tuning decisions: richening the mixture here, advancing timing there, trying a different spring or adjusting valve lash. This communication loop guided development in ways that dyno testing alone couldn't replicate.

Consider an "anatomy of a win", following a single race weekend from engine preparation to victory lane. Friday: the builder performs final assembly, carefully measuring bearing clearances and checking valve lash. The engine is started, warmed, timing and mixture adjusted on the chassis dyno. Saturday morning: practice sessions reveal a flat spot in the power delivery at mid-range RPM. The builder suspects carburetor tuning and swaps jets between sessions.

Qualifying confirms the improvement; the driver reports better throttle response and a stronger pull out of corners. The race itself: the engine runs flawlessly, maintaining oil pressure and temperature within normal ranges despite sustained high RPM. The driver manages the car well, and the combination of reliable power and smart driving delivers a victory. Sunday: teardown inspection reveals normal wear, confirming that the build was appropriately conservative for endurance rather than chasing peak numbers that might not have survived the full distance.

Think about a driver, builder, or team you admire. How much of your admiration is tied to their machinery, and how much is about their mindset and approach to problem-solving? Usually it's the latter. The best racers succeed not just because they have good equipment, but because they think clearly under pressure, communicate effectively with their teams, and make smart decisions about when to push and when to preserve equipment.

If you were building a Small Block today, whose philosophy would you follow: factory engineers emphasizing reliability and broad applicability, professional race teams pursuing maximum performance within rule constraints, or grassroots innovators in local garages finding clever solutions with limited resources? Each approach has merit. Factory philosophy emphasizes proven reliability and mass-production viability. Race team philosophy accepts higher costs and maintenance requirements in pursuit of competitive advantage. Grassroots philosophy prizes creativity and resourcefulness, finding performance through careful combinations rather than expensive components.

In this chapter, we tracked the Small Block's dominance across multiple racing arenas: NASCAR ovals where sustained high RPM and endurance mattered most, drag strips where explosive power and consistent performance earned victories, road courses where durability and broad power bands proved essential, and rugged off-road events where fundamental robustness faced the ultimate test. We saw how competition drove technical innovation: better cylinder heads flowing more air, camshaft profiles tailored to specific power-delivery goals, oiling systems managing lubrication under extreme conditions, and cooling systems handling sustained thermal loads. And we explored how racers, builders, and teams turned a production

V8 into a motorsports dynasty through talent, persistence, and accumulated knowledge.

The guiding idea of this chapter is straightforward but powerful: racing didn't just showcase the Small Block, it refined it. The same traits that made the engine a great street performer, simplicity, robustness, and adaptability, also made it a natural race platform. In turn, race-bred improvements flowed back into the aftermarket parts and build strategies that enthusiasts rely on today.

When you select a camshaft from a catalog, you're benefiting from thousands of hours of dyno testing and track time that established what works. When you choose a specific cylinder head design, you're leveraging porting knowledge accumulated through decades of competition. When you opt for a forged crankshaft over cast, you're applying lessons learned from broken parts and hard-won victories.

Next, we'll leave the formal race tracks and step fully into the world of personal expression. Chapter 7 explores how the Small Block became the canvas for custom culture, fueling hot rods, street machines, and engine swaps that turned individual creativity into rolling art without sacrificing the performance heritage you've just seen on display. The same engine that earned checkered flags also earned respect from custom builders who recognized its potential as a foundation for something uniquely their own.

When you hear the sharp crack of a Small Block at full song, whether on a short track, a drag strip, or a road course, you're listening to more than just mechanical noise. You're hearing decades of trial and error, broken parts, late nights, and hard-won victories. That racing legacy doesn't just belong to factory teams or professional drivers; it's baked into every cam profile, every cylinder head casting, and every crate engine you can order today. The checkered flags of the past are, in a very real way, under the hood of every Small Block-powered project yet to be built.

Chapter 7: The Canvas for Custom Culture

Somewhere in a two-car garage on a quiet American street in the late 1960s. The radio is playing, the door is half-open, and under a single bare lightbulb, someone is lowering a junkyard 327 into a faded '32 Ford coupe. No engineering department. No corporate budget. Just a Small Block, a pile of hand tools, and a stubborn belief that this car can be faster, louder, and more personal than anything on the showroom floor. That scene, repeated thousands of times across the country, is where the Small Block truly became the engine of custom culture.

In the last chapter, you saw how the Small Block proved itself in the harsh, competitive world of motorsports. Racing refined the engine's capability and created a deep aftermarket ecosystem. This chapter shifts from tracks to garages, from professional race teams to everyday enthusiasts, showing how those same strengths made the Small Block the go-to choice for hot rods, street machines, swaps, and today's LS-powered restomods.

By the end of this chapter, you'll understand how and why the Small Block displaced the flathead Ford as the traditional hot rod engine of choice. You'll see how the availability of parts during the muscle car era fueled a booming speed equipment industry. You'll recognize the most common performance modifications, cams, heads, intake, exhaust, displacement, and what they actually do. You'll appreciate why the "350 small-block swap" became almost universal across vehicle platforms. And you'll connect the original Small Block legacy to modern LS and restomod culture, and what that means for your own project planning.

Section 7.1: From Flatheads to Small Blocks – The Hot Rod Revolution

The Small Block Chevy didn't just join hot rodding; it redefined it, gradually replacing the Ford flathead as the preferred engine for traditional rods because of power potential, availability, and ease of modification.

Early hot rodding in the 1940s and early 1950s was dominated by Ford's flathead V8. It was simple, cheap, and for its time, relatively powerful. For a generation of builders who'd learned to work with limited resources during the Depression and wartime rationing, the flathead represented accessible performance. You could pull one from a wrecked car, strip it down on a workbench, and rebuild it with basic tools. Speed equipment companies like Edelbrock and Offenhauser had developed a solid catalog of intake manifolds, cylinder heads, and ignition components specifically for the flathead. It had become the lingua franca of American hot rodding.

But the flathead had fundamental limitations. Its side-valve design, with valves located in the block rather than in the cylinder head, created tortuous paths for intake and exhaust flow. The combustion chamber shape was inherently inefficient. As compression ratios climbed in pursuit of more power, pre-ignition and detonation became chronic problems. Cooling the areas around the exhaust valves proved difficult, leading to cracked blocks and warped heads. By the early 1950s, hot rodders were pushing the flathead to and sometimes past its practical limits.

The History of the Small Block Chevy Motor

When Chevrolet introduced the Small Block in 1955, it brought overhead valve architecture with much better breathing than the side-valve flathead design. The valves sat directly above the pistons, opening into well-shaped combustion chambers. The shorter, straighter path for intake charge and exhaust gases meant the engine could breathe more efficiently at higher RPM. Compact dimensions and light weight made it ideal for engine swaps into older, smaller chassis, a '32 Ford with a Small Block actually weighed less than the same car with a built flathead and all its heavy speed equipment. And rapidly growing factory and aftermarket support for performance parts meant you weren't starting from scratch when planning modifications.

The transition wasn't immediate. Hot rodders are conservative by nature. When you've invested time, money, and hard-won knowledge into mastering a particular engine platform, switching to something new feels risky. Early Small Block adopters were often viewed with suspicion. "Chevy guys" at predominantly Ford-based car shows sometimes found themselves defending their engine choices.

But the performance advantages were impossible to ignore. A typical transition story might go like this: a builder with a '32–'34 Ford coupe who had been running a warmed-over flathead,high-compression heads, triple carburetors, modified timing, discovers he can install a junkyard 283 or 327 with a mild cam and four-barrel carb and immediately outrun his old setup, often with better reliability. The Small Block made more power, ran cooler, lasted longer between rebuilds, and cost less to build. As hot rodders compared cost-per-horsepower and durability, the Small Block quickly emerged as the smarter choice for anyone building a car to go fast, repeatedly, and reliably.

Hot rod magazines in the late 1950s and 1960s began filling with "how-to" articles centered on Chevy small-block installations, clear evidence of the cultural shift in real time. By the mid-1960s, opening hoods at major hot rod events revealed more Small Blocks than flatheads. The engine that had started as a production powerplant for family sedans had become the new standard for traditional hot rodding.

The History of the Small Block Chevy Motor

When you think about a "traditional" hot rod today, what engine do you think of first: a flathead Ford or a Small Block Chevy, and why? If you're drawn to period-correct builds, where do you personally draw the line between authenticity and performance? Are you willing to trade originality for power and reliability? These aren't abstract questions; they shape real decisions about parts selection, budget allocation, and the ultimate character of your finished car.

Here's a practical exercise: choose one iconic hot rod era, late 1950s, early 1960s, or early 1970s. Make a short list of typical body styles used in that period (Deuce coupes, Model A roadsters, early '30s sedans), common Small Block displacements and years (265s and 283s in the late '50s, 327s in the mid-'60s, 350s by the early '70s), and typical intake and carburetor setups from that era (dual quads, three-twos, single four-barrel). Use this as a baseline if you ever plan a "period-flavored" build. Understanding what was common and what was considered radical in a given era helps you make informed choices about authenticity versus performance.

Section 7.2: Street Rods and the Transformation of Classics

As hot rodding matured into street rodding, the Small Block became the go-to powerplant for turning older cars and trucks into comfortable, reliable, high-performance drivers, bridging nostalgia and modern usability.

By the 1970s and 1980s, the hot rod scene was changing. Many enthusiasts still loved prewar and early postwar body styles, but their expectations had evolved. They wanted modern highway speeds and acceleration. They wanted better drivability, fuel economy, and starting reliability, the ability to turn a key and go, not spend twenty minutes coaxing a cold engine to life. They wanted compatibility with automatic transmissions, power steering, and air conditioning. The cars they admired from the 1940s and '50s were beautiful, but the reality of driving them daily, or even regularly, often fell short of the romance.

The History of the Small Block Chevy Motor

This shift gave birth to what became known as "street rodding," a philosophy that valued driving enjoyment as much as visual impact. Street rodders weren't trying to build the most historically accurate car or the absolute fastest quarter-mile machine. They wanted something they could drive to work, take on weekend tours, and enjoy without constant mechanical drama.

The Small Block Chevy fit this new street rod philosophy perfectly. It could be built mild or wild, depending on the intended use. A mild 350 with a relatively stock cam, modest compression, and a well-chosen carburetor or early throttle-body fuel injection delivered a smooth idle, good fuel economy, and adequate power for highway cruising, while still offering enough performance to feel genuinely quick. If you wanted more, the same basic package could accept a hotter cam, better heads, and more aggressive tuning without requiring a complete teardown.

The Small Block integrated easily with modern accessories and aftermarket front suspensions. Companies developed complete packages: engine mounts, headers designed to clear steering components, oil pans with the correct sump location for different chassis, accessory drive systems that accommodated power steering pumps and air conditioning compressors. You could order a catalog, select components by part number, and assemble a complete installation with reasonable confidence that everything would fit and work together.

And it was inexpensive compared to exotic or brand-specific alternatives. A Ford street rod builder who wanted V8 power faced either adapting a Ford engine (which meant dealing with limited aftermarket support for older Ford Y-blocks or FE-series engines) or accepting a "non-Ford" powerplant. The Small Block's ubiquity meant salvage yard cores were abundant, rebuild parts were stocked at every local parts store, and qualified mechanics were available in virtually any town.

The History of the Small Block Chevy Motor

This era solidified a pattern that continues today: take a classic shell, graft in modern running gear, usually anchored by a Small Block, and create a car you could drive across the country. A 1949–1954 Chevy or early Ford pickup running a crate 350, overdrive automatic, power disc brakes, and air conditioning became capable of long-distance road trips yet remained visually rooted in the 1950s. Street rod events such as National Street Rod Association gatherings became packed with pre-1948 vehicles, and opening almost any hood revealed a Small Block, often dressed with polished valve covers and period-style induction that nodded to traditional aesthetics while delivering modern performance.

If your goal is regular, long-distance driving in a classic vehicle, how much originality are you prepared to give up under the hood to gain comfort and reliability? Do you see a street rod as more of a "driver with style" or as a "historical artifact with upgrades"? How does that affect your engine and drivetrain choices? These questions matter because they determine not just what you buy, but how you'll actually use and enjoy your finished car.

Section 7.3: The Muscle Era and the Birth of the Speed Equipment Industry

The muscle car era didn't just sell cars; it created a massive aftermarket. The Small Block sat at the center of this explosion, driving the development of camshafts, intake manifolds, cylinder heads, and other performance parts that still shape builds today.

Through the 1960s and early 1970s, Chevrolet's own high-performance offerings, 302 Z/28s, 327s, 350 LT-1s, proved how much potential was hiding in the Small Block's architecture. These factory hot rods demonstrated that with the right combination of parts, a small-displacement V8 could deliver tire-shredding torque and high-RPM horsepower while remaining streetable. Each new performance variant from Chevrolet essentially published a blueprint that aftermarket companies could study, reverse-engineer, and improve upon.

The History of the Small Block Chevy Motor

Aftermarket companies responded aggressively. Camshaft grinders offered increasingly specialized grinds for drag racing, road racing, and street performance, each optimized for specific RPM ranges, compression ratios, and vehicle weights. Intake manifold manufacturers experimented with single-plane designs that favored top-end power, dual-plane layouts that preserved low-speed torque, and even exotic tunnel-ram configurations for all-out racing applications. Cylinder head suppliers chased airflow, refining port shapes, combustion chamber designs, and valve configurations to extract every possible horsepower.

What made the Small Block uniquely attractive for this development was its massive installed base. Because Small Blocks were everywhere, in Camaros and Corvettes, in trucks and vans, in Novas and Chevelles, investments in research and development could be spread over a huge customer base. This meant more rapid innovation, as companies could afford to experiment knowing that successful products would find eager buyers. It meant lower per-part cost compared to less common engines, making performance parts accessible to builders on modest budgets. And it meant a deeper catalog of parts at every performance and price level, from budget-friendly cast-iron pieces to exotic billet components for no-compromise race engines.

Classic catalogs from companies like Edelbrock, Holley, and Isky are filled with Small Block-specific intakes, carburetors, cams, and valvetrain components, often featuring dedicated "Chevy Small-Block" sections that run many pages. Opening one of these catalogs was like stepping into a performance buffet where you could select exactly the combination of parts that matched your goals, budget, and skill level.

The results could be dramatic, even with modest modifications. Budget backyard builds where a bone-stock 350 received a mild performance cam, a dual-plane aluminum intake, and a properly tuned four-barrel carburetor could pick up 50 or more horsepower, enough to transform a sleepy family sedan into something genuinely quick. More aggressive combinations, higher compression, ported heads, larger camshafts, and free-flowing headers could double factory power output while still maintaining reasonable street manners.

This aftermarket explosion created something else: a body of shared knowledge that made Small Block performance accessible to ordinary enthusiasts. Magazines published dyno tests comparing different combinations. Speed shops offered advice based on thousands of real-world builds. Local racers shared what worked and what didn't. Over time, certain combinations became recognized as proven formulas: the "350/350" build (350 cubic inches, 350 horsepower), the "383 stroker," the "327/302" high-winding small-cube combination. You didn't need to be an engineer to build a strong Small Block; you just needed to follow established recipes and avoid known pitfalls.

When you look at the modern aftermarket, do you see it as overwhelming or empowering? How comfortable are you navigating performance claims and marketing language? Are you more inclined to buy proven "recipe" combinations or to mix and match parts on your own? What risks and rewards come with each approach?

Here's something practical you can do right now: pick one reputable performance parts supplier, Summit Racing, Jegs, Speedway Motors, or another established vendor. Locate their Small Block Chevy section. Identify at least one "matched" top-end kit (cam, heads, intake) designed to work as a package. Note the advertised power range, RPM band, and recommended compression ratio. This will give you a concrete sense of how the industry packages Small Block performance today and help you recognize when parts are properly matched versus randomly assembled.

Section 7.4: Core Modifications – Cams, Heads, Intake, Exhaust, and Displacement

Most Small Block performance builds revolve around a familiar toolkit: camshaft choice, cylinder head upgrades, intake and exhaust flow, and sometimes increased displacement. Understanding how these elements interact is critical to avoiding mismatched, disappointing combinations.

Let's start with camshaft selection, because it's often the first modification enthusiasts consider, and the one most likely to cause problems if chosen poorly. The camshaft controls valve timing and lift, shaping the engine's torque curve and personality. A mild cam with relatively short duration and modest lift preserves good low-speed torque and smooth idle quality. A more aggressive cam with longer duration and higher lift can increase top-end power, but at the cost of reduced vacuum, rough idle, and poor low-RPM performance. Too much cam with stock heads and low compression can result in poor low-speed manners and reduced real-world performance compared to a milder, better-matched combination.

The key principle is that camshaft selection must match the rest of the engine package. A cam is not an isolated component; it's the conductor of an orchestra that includes cylinder heads, intake and exhaust systems, compression ratio, and vehicle weight. Change one element without considering the others, and the whole combination suffers.

Cylinder heads determine how well the engine breathes. Airflow and port velocity, how much air moves through the head and how fast it moves, determine maximum power potential and where in the RPM range that power arrives. Modern aftermarket heads can dramatically outperform original castings while maintaining street manners. Better port design, larger valves, improved combustion chamber shapes, and stronger materials all contribute. But this is crucial: cylinder heads must match the camshaft and intended usage. Big-port, high-flow heads designed for racing sacrifice low-speed port velocity and can actually hurt performance in a street engine that rarely sees high RPM.

Intake and exhaust systems complete the breathing package. Intake manifold design must match the intended RPM range and cam profile. Single-plane manifolds favor high-RPM power but sacrifice low-speed torque. Dual-plane manifolds maintain good low- and mid-range performance while giving up some top-end capability. Headers and a properly sized exhaust free up power by reducing backpressure and improving scavenging (the process in which exhaust pulses help draw fresh mixture into the cylinders), but the header's primary tube diameter and collector sizing must match the engine's displacement and RPM range. Bigger is not always better; too-large tubes reduce velocity and hurt performance.

Finally, displacement increases offer a straightforward path to more torque. Stroker combinations, created by installing a crankshaft with a longer stroke, raise displacement and torque production. A popular combination uses a 350 block with a longer-stroke crank to create 383 cubic inches. The extra displacement makes the engine more forgiving of larger cams and airflow improvements, delivering strong low-end torque even with parts that would otherwise work best at higher RPM.

The central principle tying all of this together: the best Small Block builds use matched components designed to work together, not a random assortment of "biggest" or "most extreme" parts. Here's what a well-matched street build might look like: 350 cubic inches with 9.5:1 compression, a mild hydraulic flat-tappet cam with moderate duration (around 210–220 degrees at .050-inch lift), an aluminum dual-plane intake with a properly sized carburetor (600–650 cfm) or entry-level fuel injection, modern aftermarket heads with good low- to mid-lift flow characteristics, 1 5/8-inch primary headers, and a free-flowing dual exhaust. The result: strong off-idle torque, good street manners, easy drivability, and enough power to feel genuinely quick without compromises in daily use.

Contrast that with a mismatched combination: stock low-compression 305, a very large cam intended for high-RPM race use, and a single-plane intake on a heavy street car with tall rear gears. The result: soft bottom end, poor idle quality, disappointing real-world acceleration, and an engine that only comes alive at RPM ranges you'll never use on the street.

The History of the Small Block Chevy Motor

When you think about your ideal build, are you imagining peak horsepower numbers or usable torque in the way you actually drive? How does that change your parts priorities? Have you ever been tempted by "race" parts for a streetcar? What trade-offs are you truly willing to live with day-to-day?

Section 7.5: The 350 Swap – A Universal Powerplant

The Small Block 350 became the de facto "default engine swap" for decades. Its combination of size, weight, availability, and aftermarket support made it a practical choice for everything from classic cars and trucks to unconventional platforms.

The 350-cubic-inch displacement hit a sweet spot, balancing multiple competing priorities. It offered enough displacement to deliver serious torque on the street, torque you could feel launching from stoplights and merging onto highways. It was common in trucks and passenger cars from the late 1960s through the 1990s, feeding an enormous used-and-core market. You could walk into almost any salvage yard in America and find multiple 350 cores in varying conditions at reasonable prices. And it was supported by GM's crate engine programs, which offered new, warrantied complete engines in configurations ranging from bone-stock rebuilds to serious performance packages.

The engine's physical characteristics helped enable its swap popularity. Compact external dimensions relative to displacement meant it could fit into engine bays originally designed for much smaller powerplants, flexible mounting options, with adaptation kits available for virtually every popular chassis, and simplified installations. Broad transmission compatibility meant you could mate the engine to everything from vintage three-speeds to modern overdrive automatics. Oil pans were available in front-sump, rear-sump, and center-sump configurations to clear different chassis components.

The cultural outcome of all these practical advantages: "Throw a 350 in it" became a standard answer to almost any engine problem, from worn-out original powerplants to ambitious performance

The History of the Small Block Chevy Motor

upgrades. That six-cylinder pickup that's tired and slow? Drop in a 350. That vintage Jeep that can't keep up with modern traffic? Install a 350. That foreign sports car with a temperamental engine and no parts support? Swap in a 350. The solution became almost reflexive.

Common swap scenarios played out in garages across the country. Replacing a tired six-cylinder in a 1960s or 1970s Chevy with a mild 350 crate engine was nearly plug-and-play; existing motor mounts often worked, transmissions bolted up, accessories transferred, and the whole job could be completed in a weekend. Installing a 350 into non-GM platforms, Jeeps, older Fords, and various imports required more fabrication but was well-supported by widely available adapter kits and engine mounts. Entire cottage industries grew up around specific swaps: Small Block Chevys into early Broncos, into Datsun Z-cars, into Jeep CJs and Wranglers.

Many "recipe" builds in magazines and books specified a 350 purely for its predictability. Technical writers could assume readers would find parts and support almost anywhere. That universality created a self-reinforcing cycle: the more popular the 350 became for swaps, the more swap parts became available, which made it even more popular for future swaps.

How important is parts availability and local expertise to you when choosing an engine? Would you sacrifice uniqueness for easier sourcing and support? Do you see the 350 swap as "cliché" or as a smart, proven solution? How does that perception influence your own build decisions? These aren't rhetorical questions; they touch on fundamental aspects of what you value in a project and how you'll experience building and owning the finished vehicle.

Here's something concrete you can research right now: for the vehicle you own or hope to build, quickly investigate whether swap kits or mounts exist specifically for a Small Block 350. Look for which oil pan is commonly used, which header configurations work, which accessory drive setups are typical, and whether there are known clearance issues with steering boxes, crossmembers, or hood lines. Capture what you find in a simple "swap feasibility" note you can refer to later. These five or ten minutes of research can save you weeks of frustration and hundreds of dollars in wrong parts.

Section 7.6: From Carburetors to Coil Packs – LS Engines and Modern Restomod Culture

Modern LS-series engines extended the Small Block philosophy into the fuel-injected era, becoming the new backbone of restomod culture while preserving many of the same values: compact size, broad availability, and exceptional performance per dollar.

While LS engines are technically a new generation, different block architecture, different heads, different everything when you look closely, they carry forward essential Small Block traits that matter to enthusiasts. Compact, efficient packaging makes them suitable for swaps into vehicles never designed for them. Robust bottom ends, capable of handling significant power increases, respond well to modifications. Strong aftermarket and OEM crate programs provide parts support that rivals, and in some ways exceeds, what was available for earlier Small Blocks.

Restomod builders increasingly choose LS powerplants because factory fuel injection and electronic ignition improve drivability and reliability in ways that carburetors and distributors simply cannot match. Modern engine management systems adjust fuel delivery and timing constantly, compensating for temperature, altitude, and driving conditions without requiring manual intervention. Stock engines deliver power levels that previously required extensive modification. A bone-stock truck-spec 5.3L LS produces more usable power than most built first-generation Small Blocks while returning better fuel economy and requiring less maintenance.

Modern engine management also allows tuning for emissions and efficiency without sacrificing performance. In many jurisdictions, passing emissions testing with a carbureted engine built to modern power levels ranges from difficult to impossible. LS swaps with proper programming can meet emissions standards while delivering hundreds of horsepower. For builders who want to drive their cars

The History of the Small Block Chevy Motor

regularly without legal complications, that capability alone often justifies choosing an LS over a traditional Small Block.

This evolution means the "Small Block legacy" is no longer limited to carbureted Gen I engines; its spirit lives on in every classic shell running a modern, efficient V8 under the hood. A 1969 Camaro or 1970 Chevelle with an LS3 crate engine, six-speed manual or modern automatic with overdrive, and updated suspension and brakes preserves the visual and emotional feel of a muscle-era icon while performing like a modern sports car. It starts instantly in any weather, idles smoothly, delivers crisp throttle response, and can be driven cross-country with confidence.

Builders who once reflexively chose a carbureted 350 now default to 5.3L or 6.0L truck-derived LS swaps, attracted by low donor cost (used LS truck engines are abundant and relatively inexpensive) and abundant tuning support. The same deep aftermarket that supported earlier Small Blocks now offers LS swap kits for countless applications, including engine mounts, oil pans, headers, wiring harnesses, and complete fuel systems. Online forums provide detailed build threads and troubleshooting advice. The knowledge base that made traditional Small Block swaps accessible has transferred to the LS platform.

Where do you personally fall on the spectrum between "old-school carb and points" and "modern EFI and electronics"? What experiences shaped that preference? If you're building a car to keep for decades, does the long-term serviceability and efficiency of an LS-type engine change your thinking compared to a classic Small Block? There's no wrong answer, but understanding your own priorities helps ensure your finished project matches your actual needs and preferences.

In this chapter, you've traced how the Small Block Chevy became the backbone of custom culture. You saw it displace the flathead in traditional hot rods, power the rise of street rodding, and drive the growth of the performance aftermarket. You examined core modifications and the logic behind them, explored why the 350 swap became almost universal, and followed the lineage into modern LS-powered restomods.

The central idea running through all of this is adaptability. The Small Block, and later its LS descendants, became the mechanical "language" through which enthusiasts customize, personalize, and modernize their vehicles. Understanding that language gives you both historical appreciation and practical guidance for your own builds. The engine's success in custom culture wasn't accidental or purely the result of good marketing. It emerged from real engineering excellence, broad availability, and the accumulated wisdom of millions of enthusiasts who discovered what worked and shared that knowledge freely.

So far, most of our story has unfolded on American streets, strips, and show fields. In the next chapter, you'll zoom out to a global perspective, following the Small Block as it crosses oceans, powers foreign vehicles, and shapes automotive culture in places far from Detroit. You'll see how the same qualities that fueled American custom culture resonated with builders and manufacturers around the world, and how the Small Block's influence extended far beyond the borders of the country where it was born.

The History of the Small Block Chevy Motor

Chapter 8: The Small Block Goes Global

Envision this: you're standing on a crowded street corner in São Paulo, watching a battered taxi idle at the curb, its exhaust note rumbling against the concrete and glass. A few months later, you're halfway around the world at Mount Panorama, where the howl of a V8 echoes off the hills as a touring car crests Skyline. Half a planet apart, completely different missions, yet the heartbeat under both hoods traces back to the same Small Block Chevy architecture.

By now, you've seen how the Small Block conquered American streets, dominated competition, and became the default canvas of custom culture. This chapter widens the lens. We're following the same engine architecture as it crosses oceans, adapts to new markets, and influences foreign manufacturers and engineers in ways that even surprised Chevrolet.

Understanding this global journey helps explain why parts, designs, and ideas surrounding the Small Block didn't stay confined to the United States. They bounced back and forth across borders, shaping not only what people drove, but how engineers around the world thought about performance and durability. The Small Block's story isn't just American, it's genuinely global, and that reach reveals something fundamental about what made this engine truly revolutionary.

Section 8.1: Exporting an American Heartbeat

The Small Block's global story doesn't begin with foreign manufacturers copying it; it starts with Chevrolet and General Motors deliberately shipping it abroad. First as a powerplant in export cars and trucks, then as an industrial workhorse, and finally as a completely knocked down (CKD) assembly component that would seed local familiarity with the design among technicians who'd never seen Detroit.

As Chevrolet realized how successful the Small Block was at home, GM's overseas divisions recognized the same opportunity: here was a compact, relatively light, torquey V8 that could be adapted to different fuels, climates, and duty cycles. The engine's modular architecture and tolerance for variations in fuel quality made it attractive in markets where infrastructure wasn't as consistent as in North America. In places with limited service networks, parts commonality and simple service procedures weren't just conveniences; they were survival traits.

Export programs shipped complete cars and trucks, as well as separate engine assemblies, to Europe, South America, Africa, and Asia. In some cases, these engines were installed in Chevrolet-branded vehicles; in others, they powered local GM brands and even non-GM applications. Marine variants found their way into fishing boats off the coast of Chile. Industrial versions drove generators in remote African mining operations and irrigation pumps in Southeast Asian rice paddies. Wherever robust service networks were scarce but basic mechanical skills were common, the Small Block's straightforward design became an asset worth more than raw horsepower numbers.

The History of the Small Block Chevy Motor

Chevrolet trucks sold in export markets, particularly in Latin America and the Middle East, often carried Small Block V8s as optional or standard powerplants, giving local operators access to American V8 torque for hauling and commercial work. These weren't prestige vehicles. They were workhorses, expected to haul cargo over terrible roads, operate in dust and heat that would challenge far more sophisticated engines, and keep running with minimal maintenance. The fact that they did exactly that, year after year, built the Small Block's reputation in markets where reliability mattered more than performance.

CKD kits played a particularly important role. To avoid import tariffs, vehicles were shipped in pieces and assembled locally. These kits often included Small Block engines, which meant local assembly plants needed trained technicians who understood the engine's architecture. Those technicians talked to local mechanics. Local mechanics talked to enthusiasts. And suddenly, in markets thousands of miles from Detroit, people understood how to build, rebuild, and modify Small Block Chevys, not because of advertising campaigns, but because the engines were physically present and practically unavoidable.

When you see a famous engine outside its home market, what do you think allowed it to survive that journey? Was it marketing, engineering, or sheer practicality? In the Small Block's case, marketing helped, but the engine's survival came down to something more fundamental: it was useful. Not exotic, not prestigious in the traditional sense, just reliably, adaptably useful in ways that transcended cultural and economic boundaries.

If you work with or restore vehicles in a non-U.S. market, take a moment to notice any traces of American V8 influence in local fleets or older work vehicles. The connections might surprise you. That generator at a remote radio station, the marine engine in a harbor workboat, the faded pickup truck still hauling supplies in rural areas, chances are, more than a few trace their lineage back to Chevrolet's compact V8.

Section 8.2: British Sports Cars and European Hybrids

Throughout Europe, but especially in the United Kingdom, a fascinating hybrid culture emerged during the 1960s and 1970s: light, nimble European sports cars powered by American Small-Block V8s. This wasn't sacrilege to everyone involved. For many builders and buyers, it was the perfect marriage, European chassis finesse meeting American torque in combinations that delivered performance neither could achieve alone.

The Small Block's compact external dimensions made it uniquely suitable for engine swaps into cars never designed for V8 power. While big-block engines were simply too bulky and heavy for most European chassis, the Small Block could slip into many engine bays originally intended for inline-fours or sixes, with careful engineering and a willingness to modify mounting points, cooling systems, and transmissions.

British and European manufacturers seeking reliable, powerful, and relatively affordable engines for low-volume sports cars often lacked the resources to develop their own high-performance V8s from scratch. Development costs for a clean-sheet engine design were prohibitive for small manufacturers. Buying proven American V8s off the shelf was a practical solution that also delivered impressive results. The Small Block brought not just power, but parts availability and a global support network that no boutique European engine could match.

This created a wave of Anglo-American and Euro-American hybrids in which the Small Block became an acceptable, even desirable, foreign transplant. Once buyers realized they could get readily available parts and tuning knowledge worldwide, the engine's American origins became less important than its proven capabilities. A Small Block-powered TVR or Ginetta wasn't pretending to be American; it was using American technology to achieve distinctly British performance goals.

Small-Block-powered specials and kit cars proliferated across the UK, where builders dropped Chevy V8s into everything from Ford Cortinas to purpose-built tube-frame chassis designed around American power. These weren't all high-dollar builds. Many were garage projects undertaken by enthusiasts who understood that a used Small Block and a weekend's worth of fabrication work could transform an ordinary car into something genuinely quick.

European race and rally specials experimented with American V8s, drawn by their durability and tuning potential for endurance events where mechanical resilience mattered as much as outright speed. When you're facing a 24-hour race or a multi-day rally, the ability to run at high load without catastrophic failure becomes paramount. The Small Block's reputation for exactly that kind of durability made it attractive to team managers who needed to finish, not just lead the first lap.

Specialist manufacturers and tuning houses across Europe offered conversion packages, engine mounts, and cooling solutions tailored specifically to installing Small Blocks into European platforms. Some of these companies became significant players in their own right, developing expertise and product lines that served an international market. Their catalogs documented which transmissions mated easily to Small Block bellhousing patterns, which radiators fit which chassis configurations, and which exhaust manifolds cleared steering components in tight engine bays.

How do you feel about cross-cultural engine swaps? Does dropping an American V8 into a European chassis create something exciting, or does it somehow dilute the purity of the original design? There's no single answer; it depends on what you value. If originality and manufacturer intent matter most, hybrids can feel like compromises. But if driving experience and practical performance take priority, a well-executed V8 conversion can make a light European chassis come alive in ways the original four-cylinder or six never could.

If you've ever driven or ridden in a lightweight car with a torquey engine, you understand how dramatically that combination changes the vehicle's character compared to its stock form. The throttle response sharpens. The rear suspension works harder. The whole machine feels more alert, more eager, occasionally more difficult, but rarely boring.

Section 8.3: Brazil, South America, and Local Production

In South America, especially Brazil, the Small Block didn't just arrive as an imported curiosity. It became a locally produced, deeply integrated part of the automotive ecosystem, powering both passenger vehicles and workhorses tailored to regional needs. This wasn't re-badging or simple assembly work. This was adaptation, refinement, and in some cases genuine innovation driven by local conditions that Chevrolet's engineers in Michigan never had to consider.

Brazil's combination of import restrictions, local content rules, and a growing middle class pushed GM do Brasil to develop and manufacture engines locally rather than import them fully built. The Small Block architecture, already proven and flexible, was a natural foundation. But local production meant more than just setting up assembly lines. It meant adapting engines to available fuels, including ethanol blends, long before they became common in North America, and tuning them for ambient temperatures and loads that Brazilian customers faced daily.

These weren't minor tweaks. Brazilian Small Blocks often featured different compression ratios to handle lower-octane fuel or high-ethanol content. Cooling systems were modified to manage tropical heat. Emissions equipment evolved to meet local regulations that sometimes diverged from U.S. standards. In essence, Brazilian engineers were solving the same problems their American counterparts faced, but with different constraints and different solutions.

The History of the Small Block Chevy Motor

As these engines spread through fleets of taxis, trucks, and utility vehicles, Brazilian mechanics and tuners developed their own expertise around the Small Block. This wasn't just an extension of American hot rodding culture translated into Portuguese. It was an independent performance scene with its own priorities, techniques, and innovations, clearly connected to the U.S. tradition but distinct in meaningful ways.

GM C/K trucks and their Brazilian derivatives used Small Block-based V8s for commercial work, cargo hauling, and rural transport. To someone familiar with American trucks, these vehicles looked recognizable, with similar body lines and similar proportions. But under the hood, specifications often diverged. The Brazilian market demanded different torque curves, durability standards, and maintenance intervals based on how and where these trucks actually operated.

Locally specified engines designed to handle ethanol or low-octane fuel showcased the Small Block's ability to be retuned and reconfigured for different combustion characteristics. Ethanol burns cooler but requires richer fuel mixtures and different ignition timing than gasoline. Brazilian engineers worked out these calibrations decades ago, building expertise that would later inform global discussions about alternative fuels and flex-fuel technology.

Perhaps most interestingly, a Brazilian drag racing and street performance scene built itself around reworked Small Blocks, complete with region-specific parts suppliers, machine shops, and tuning practices. Some of these innovations eventually fed back to North American enthusiasts via online communities and international events. A camshaft profile developed for Brazilian ethanol fuel might work beautifully in a U.S. engine running E85. A cylinder head design optimized for low-octane durability might offer insights for builders dealing with pump gas limitations. The exchange wasn't one-way; it was circular, with ideas and solutions flowing in multiple directions.

What happens to an engine's identity when it's manufactured, tuned, and raced thousands of miles away from where it was designed? Is it still "American," or does it become something new? The question matters less than the reality: the Brazilian Small Block was both American and distinctly Brazilian, shaped by the same core architecture but refined by different hands for different purposes.

If you're in a market with local variants of global engines, consider how often those regional changes actually make the engine better suited to your real-world use. The modifications weren't arbitrary; they reflected a genuine understanding of local conditions, local fuels, and local driving patterns. That knowledge has value wherever similar conditions exist.

Section 8.4: The Australian Connection: Holden and Beyond

In Australia, the Small Block didn't just show up; it helped shape an entire performance culture. Through Holden and related GM programs, Small Block-derived power helped define Australian muscle cars, touring car racing, and the country's distinctive V8 identity. To understand Australian automotive culture without understanding the Small Block's role would be like trying to explain American hot rodding without mentioning Ford flatheads. The connection runs that deep.

Australia's vast distances, harsh conditions, and strong appetite for rear-wheel-drive sedans and utes made torque and durability non-negotiable. When your daily commute might involve hundreds of kilometers of highway, when summer temperatures regularly exceed 40°C (104°F), when dust and rough roads are normal rather than exceptional, your engine needs to be fundamentally robust, not just powerful. The Small Block's core strengths matched these requirements remarkably well.

Holden and other GM affiliates adapted Small Block architecture and philosophy into their own engines and vehicle platforms.

The History of the Small Block Chevy Motor

Sometimes this meant direct imports of American-built engines. Sometimes it involved local derivatives or parallel designs inspired by the same engineering principles. The result was a family of V8s that felt American in character but Australian in execution, engines that reflected both their Chevrolet heritage and the specific demands of Australian motoring.

On the track, the Small Block and its descendants powered cars that became legends at events like the Bathurst 1000. This wasn't peripheral involvement; Chevy-derived V8s became central characters in motorsport battles that captivated fans and shaped brand loyalties for generations. When Holden and Ford battled at Mount Panorama, they weren't just racing for trophies. They were establishing cultural narratives about performance, reliability, and national identity that persist to this day.

Holden performance sedans and coupes used Small Block-based V8s or related GM V8 architectures, often tuned specifically for Australian fuel and road conditions. The iconic Holden Monaro, for instance, carried V8 power that traced its lineage directly to the Small Block family, but with specifications reflecting Australian engineering input and local market demands. These weren't badge-engineered American cars with right-hand drive tacked on; they were genuinely Australian machines built around American engine architecture.

Touring car and sedan racing programs drove both engineering innovation and brand loyalty. When fans watched V8 Supercars thunder around Bathurst or other circuits, they were seeing competition that pushed engine development in real time. Lessons learned on Sunday influenced Monday morning engineering discussions, creating a tight feedback loop between competition and production that benefited both.

Australian kit cars, hot rods, and engine swaps embraced the Small Block just as naturally as American builders did. Local parts manufacturers and machine shops developed Australian-specific solutions to Australian-specific problems, cooling systems for extreme heat, air filtration for dusty conditions, and tuning strategies for local fuel formulations. The Small Block's flexibility allowed all of this variation while maintaining the core simplicity that made it attractive in the first place.

Why do you think Australia, in particular, bonded so strongly with V8 culture? How much of that is geography and infrastructure, and how much is national personality? The answer probably involves both. Long distances favor torquey engines that can cruise efficiently at highway speeds for hours. Hot, dusty conditions favor simple, robust designs that don't require constant fiddling. But there's also something culturally resonant about V8 power in the Australian context, a connection to ideas about capability, self-reliance, and performance that goes beyond pure engineering.

If you've watched Australian touring car races, how does the sound and character of those V8s compare in your mind to American stock car or Trans-Am racing? The engine architecture might be similar, but the racing style, the circuits, and the presentation are all distinctly Australian. The Small Block adapted to that context, proving once again that good engineering isn't rigidly specific. It's adaptable.

Section 8.5: Global Motorsports: From Le Mans to Bathurst

The Small Block's racing legacy wasn't confined to American ovals and drag strips. Its reliability, compactness, and tuning range made it a natural candidate for international endurance races, touring car battles, and specialized regional series across the globe. Wherever motorsport existed, the Small Block eventually appeared, sometimes in factory-backed efforts, often in privateer hands, always proving that American pushrod V8 technology could compete with anything the world offered.

Endurance racing in Europe and elsewhere tends to punish fragile engines mercilessly. Teams don't just need speed; they need power plants that can run at high load for hours, even days, without catastrophic failure. The Small Block's sturdy bottom end and proven oiling solutions made it a logical candidate for exactly this kind of competition. When Corvette Racing campaigned at Le Mans, they weren't gambling on an untested engine. They were leveraging decades of durability development dating back to 1955.

In many series, regulations favored production-based engines or at least production-derived architecture. The Small Block's enormous production volume and wide range of displacements allowed teams to tailor builds to different rulebooks,displacement-limited classes, power-to-weight formulas, and fuel economy requirements. Need a 5.0-liter for Trans-Am? Build a 305. Need maximum displacement for unrestricted classes? Stroke a 400. The flexibility was tactical, not just technical.

As the engine ventured into foreign racing series, local builders began making their own refinements, cylinder heads, cam profiles, and lubrication strategies, optimized for specific circuits and climates. A road-racing engine built for Nürburgring faced different demands than one built for Bathurst, which in turn faced different demands from one built for Sebring. Yet all could be traced back to the same basic design. That feedback loop, in turn, informed the global aftermarket, as innovations developed for competition eventually found their way into street parts catalogs.

Corvettes and other GM entries at Le Mans and European endurance events demonstrated that Small Block-based V8s could deliver both speed and durability against sophisticated European rivals. When a Corvette C5-R or C6.R finished at Le Mans, not just finished, but won its class or even took overall victories, it validated decades of pushrod V8 development in the most demanding test environment motorsport offers. Critics who dismissed American V8s as crude or outdated had to reckon with lap times and reliability records that spoke louder than theory.

Australian endurance races like Bathurst saw Chevy-derived V8s become central characters in battles that captivated fans and shaped brand loyalties for generations. The mountain circuit demanded engines that could deliver power reliably through Conrod Straight, then pull hard out of tight corners, then cool adequately on the long climb back up the hill, lap after lap, hour after hour. The Small Block's thermal management and mechanical robustness made it well-suited to exactly this kind of punishment.

Regional GT and touring series in Europe, South America, and elsewhere saw privateer teams choose Small Block-based engines specifically because parts, knowledge, and support were widely available. When you're a small team operating on a modest budget, you can't afford exotic engines that require factory support or unobtainable components. The Small Block offered a different value proposition: proven performance backed by a global infrastructure of parts and expertise. A broken valve spring at a circuit in Argentina? Someone nearby probably had spares. A head gasket failure in Italy? Local machine shops understood the specifications. That practical support network mattered as much as horsepower numbers.

When you think of a "global" racing engine, what comes to mind first, something exotic and rare, or a robust, mass-produced design refined by thousands of builders worldwide? Both have their place, but the Small Block proved that the latter approach could compete at the highest levels. The engine's global success came not from mystique or limited availability, but from being so thoroughly understood and so widely supported that teams anywhere could build competitive packages.

How might the pressure of 24-hour races or long-distance events push engine development in ways that ordinary road use never would? Endurance racing is an accelerated lifetime test. An engine that survives Le Mans has faced thermal cycles, vibration, sustained high loads, and stress concentrations that might take years to encounter on the street. Problems that might emerge after 100,000 road miles show up after 24 hours at racing speeds. The development feedback from this kind of testing benefits everyone, racers and street enthusiasts alike.

Section 8.6: A Worldwide Aftermarket and Tuning Community

As the Small Block spread around the world, it didn't just export hardware; it helped seed a global ecosystem of tuners, machine shops, and parts suppliers who interpreted and improved the engine through their own regional lenses. What emerged wasn't a single, monolithic Small Block culture, but rather a network of interconnected communities, each contributing insights and innovations that enriched the whole.

The core reasons that made the Small Block attractive to American hot rodders, interchangeable parts, simple architecture, and a vast catalog of components, translated naturally into other markets. Once a few builders in a region gained experience and proved that Small Block swaps or builds were viable, knowledge spread quickly. The engine's inherent simplicity meant that fundamental techniques worked similarly across geographies. A valve job performed correctly in Japan followed the same principles as one performed correctly in Sweden or Chile.

Yet import constraints and high shipping costs often pushed international enthusiasts to improvise in ways American builders rarely needed to consider. When ordering parts from the United States meant weeks of shipping time and potentially crippling import duties, local solutions became necessary. Builders machined local components to work with Small Block dimensions. They adapted non-Chevy parts when genuine GM pieces were unavailable or prohibitively expensive. They sometimes even cast region-specific heads and manifolds designed around local manufacturing capabilities and material availability.

This forced creativity often produced genuinely innovative solutions. A European shop developing high-flow cylinder heads optimized for road racing and high-RPM use might approach the problem differently from an American drag-racing shop focused on low-end torque and nitrous compatibility. Neither approach was inherently superior; they simply reflected different priorities and different racing cultures. But when these different approaches became visible to the global enthusiast community, everyone benefited from the expanded toolkit of ideas.

The History of the Small Block Chevy Motor

With the rise of the internet in the 1990s and 2000s, this once-local experimentation became visible worldwide almost instantly. Techniques or part designs tested in one country were adopted or refined elsewhere, creating a truly global development loop around a mid-century American engine design. A forum post from Brazil describing a budget stroker combination might inspire a builder in Poland. A dyno chart from Australia showing results with a particular camshaft might inform enthusiasts in South Africa's choices. The technology hadn't changed; the Small Block architecture remained fundamentally the same, but the collective knowledge about what could be done with that architecture expanded exponentially.

European companies offered high-flow Small Block cylinder heads and intake manifolds optimized for road racing and high-RPM use, often with port configurations and valve sizing that reflected priorities different from American drag-focused designs. These weren't just copies of U.S. parts with metric bolts; they represented genuine development work aimed at different applications. A head designed for sustained high-RPM operation on an endurance circuit emphasizes cooling, durability, and consistent airflow across a broad RPM range. A head designed for drag racing might prioritize peak flow at specific engine speeds, accepting compromises that wouldn't work in longer events.

South American and Australian tuners developed budget-friendly stroker combinations and forced-induction setups tailored to local fuel quality and economic realities. When premium racing fuel costs twice what it does in the United States, or isn't available at all, you develop different strategies. Lower compression ratios, conservative timing curves, and boost levels that work with available fuel become more important than chasing maximum numbers. These practical constraints drove innovation that often proved useful far beyond their original context.

International online forums became spaces where builders in different hemispheres shared dyno charts, camshaft profiles, and workarounds for sourcing parts that might be rare or expensive in their home markets. The conversations revealed both universal truths and regional variations. Certain combinations worked well everywhere. Others needed adjustment for local conditions. But the collective knowledge base grew richer because people from different backgrounds contributed different perspectives on the same engine family.

How has easy access to global information changed the way people modify and tune engines? Do you see more convergence toward "best practices," or more creative divergence as builders explore different paths? The answer is probably both. Certain fundamentals, proper machining tolerances, correct assembly procedures, and sound tuning principles have converged as information spread. But the applications and priorities have diverged, as builders apply those fundamentals to their own unique situations and goals.

If you build or plan to build a Small Block, how much of your parts list or strategy has been shaped by ideas that originated in other countries, even if you don't always realize it? That cylinder head design might reflect European racing development. That stroker combination might trace back to Australian street machines. That tuning approach for E85 might have roots in Brazilian ethanol experience. The global nature of Small Block knowledge isn't always obvious, but it's deeply embedded in how we think about the engine today.

Section 8.7: Cultural and Engineering Cross-Pollination

The Small Block's global spread wasn't a one-way export of American ideas. As different countries adopted and adapted the engine, they also sent concepts, expectations, and engineering lessons back into the broader automotive world. This cross-pollination, technical, cultural, and philosophical, enriched both the engine itself and how people around the world thought about accessible performance.

Engineers and tuners in regions with strict fuel, emissions, or durability constraints often found creative ways to extract performance without compromising reliability. Those solutions sometimes influenced later factory developments or aftermarket products worldwide. When Brazilian engineers worked out how to calibrate Small Blocks for high-ethanol fuel in the 1980s, they were decades ahead of the flex-fuel wave that would later sweep through North America. When European tuners developed cylinder heads optimized for sustained high-RPM operation on endurance circuits, they were solving problems that would later matter to American road racers facing increasingly competitive series.

The influence wasn't always direct or immediate. An innovation developed in one market might take years to migrate elsewhere. But the underlying principle held: good ideas travel, especially when they solve real problems. The Small Block's global presence ensured that innovations developed anywhere could potentially benefit builders everywhere.

Culturally, the presence of an American V8 in foreign cars changed perceptions on both sides. In some places, it symbolized aspiration and connection to a broader car culture, American performance mystique crossing borders. In others, it became simply a practical, proven choice rather than an exotic indulgence. The Small Block was powerful enough to be desirable, but common enough to be accessible. That combination allowed it to transcend its country of origin and become something larger: a shared reference point for enthusiasts regardless of their background.

Over time, the Small Block became a sort of common technical language. An engineer in Australia, a tuner in Brazil, and a hot rodder in Germany could all discuss bore, stroke, heads, cams, and compression around this shared reference point, even as they pursued different goals. The specifics might vary, different displacement preferences, different intended uses, different performance priorities, but the fundamental vocabulary remained consistent. This shared language facilitated knowledge exchange in ways that would have been impossible with more fragmented engine families.

Global emissions and fuel economy pressures drove the development of new camshaft designs, fuel system strategies, and combustion chamber refinements that later influenced GM's production engines and aftermarket offerings worldwide. When European regulations demanded cleaner combustion or better fuel economy, engineers developed solutions within the Small Block architecture rather than abandoning it. Those solutions, improved chamber shapes, more efficient cam profiles, and better fuel atomization often proved valuable in other markets facing similar or different constraints.

International events, track days, hillclimbs, and drag meets became venues where Small Block-powered cars from different countries competed side by side, each reflecting the priorities and styles of their builders. A Brazilian street machine built for ethanol and tight autocross courses looked and behaved differently than an Australian touring sedan built for highway speeds and Bathurst endurance. Yet both were recognizably Small Block-powered, sharing fundamental architecture while expressing different visions of what that architecture could achieve.

Technical articles and books produced outside the United States began treating the Small Block as a standard teaching platform for engine theory, spreading its influence into formal and informal education worldwide. When engineering students in Europe or mechanics-in-training in South America learned the principles of internal combustion, many did so using the Small Block architecture as the reference example. This reinforced the engine's position not just as a performance option, but as a baseline for understanding how engines work.

When you think about "American" or "European" engineering, how much of that distinction still holds up in a world where ideas, parts, and engines cross borders constantly? The categories still have meaning; different traditions emphasize different values and offer different solutions to similar problems. But the boundaries are far more porous than they once were. The Small Block's global journey exemplifies this porosity. It remained recognizably American in character, with pushrod V8 simplicity, cubic-inch displacement, and torque-focused performance, yet it absorbed influences and ideas from everywhere it went.

What have you personally learned, directly or indirectly, from builders in other countries, whether through videos, articles, or conversations? If you're engaged with Small Block culture at all, the answer is probably "more than you realize." That technique you picked up from a YouTube video might have originated in Australia. That part's recommendation from a forum might reflect European racing experience. That fuel-tuning strategy might trace back to South American expertise in alternative fuels. The knowledge flows in all directions now, creating a genuinely global community around a fundamentally simple engine.

In this chapter, we followed the Small Block from an export power plant to a globally embedded engine family. We saw how it powered work trucks and taxis in South America, slipped into European sports cars, became a cornerstone of Australian performance culture, and proved itself in international motorsports from Le Mans to Bathurst. We examined how a worldwide aftermarket and tuning community grew around the engine, fostering a genuine two-way exchange of ideas, technical, cultural, and philosophical, about what an accessible performance engine should be.

The guiding theme of this chapter, global reach and adaptation, circles back to the Small Block's original strengths: compactness, simplicity, durability, and tunability. Those qualities made it not just exportable, but adoptable by wildly different markets and cultures. An engine that worked well in American conditions could be tuned to work well in Brazilian heat, Australian dust, European fuel regulations, and countless other contexts. That adaptability wasn't accidental. It flowed from design decisions made in 1954 that prioritized fundamental soundness over specialized optimization.

In the context of the whole book, the Small Block's global journey reinforces a central idea: when an engineering solution is fundamentally sound and genuinely usable, it transcends borders. It stops being just "Chevrolet's V8" and becomes part of a shared mechanical heritage. The Small Block achieved this transformation not through marketing or force of will, but by simply being good enough and flexible enough to serve different needs in different places without losing its essential character.

Next, we'll shift our focus from geography to symbolism. If this chapter showed how the Small Block became physically present around the world, the next chapter explores how its sound, image, and mythology seeped into movies, music, television, and digital media. In Chapter 9, we'll see how the engine evolved from a piece of machinery into a cultural icon, a symbol that means "power," "freedom," or "rebellion" even to people who have never owned a car, let alone turned a wrench on a Small Block.

Stand at almost any major car meet in the world, close your eyes, and listen. Sooner or later, you'll hear that familiar V8 cadence, sometimes muffled in a worn-out taxi, sometimes sharpened in a race car, sometimes echoing off city walls half a planet away from Detroit. That sound is more than an echo of American engineering. It's a reminder that good ideas travel, adapt, and take root wherever people care about machines. The Small Block Chevy didn't just go global in shipping containers and dealer catalogs; it did so in the imaginations and workshops of enthusiasts everywhere, setting the stage for the cultural story we explore next.

The History of the Small Block Chevy Motor

Chapter 9: The Pop Culture Engine

How did a mass-produced Chevrolet V8 evolve from a piece of hardware into a cultural symbol, one whose sound, silhouette, and reputation show up everywhere from movie screens to music lyrics and video games? In the previous chapter, you saw how the Small Block crossed borders and influenced automotive culture around the globe. Now we shift from geography to symbolism. This chapter explores how the engine stepped beyond engineering, racing, and manufacturing to occupy a new role: a shorthand for freedom, rebellion, nostalgia, and American performance in popular media.

By the end of this journey through celluloid, sound waves, and digital realms, you'll understand how film and television turned Small Block–powered cars into cultural icons. You'll see why musicians, writers, and journalists repeatedly invoke the Small Block when they talk about speed, freedom, and identity. You'll discover how video games and digital media preserved and amplified the Small Block's legend for new generations. And perhaps most importantly, you'll appreciate why certain Small Block–powered vehicles and engines command high collectibility and museum interest, often for cultural reasons as much as technical ones.

The History of the Small Block Chevy Motor

Section 9.1: Hollywood's Love Affair with the Small Block

Hollywood didn't just use Small Block–powered cars as props; it used them as visual and auditory shorthand for attitude, era, and personality, from the cruising culture of the 1960s and '70s to the high-stakes heists and street racing of modern action films.

Think about it for a moment: when directors needed to communicate a character's rebellious streak, working-class authenticity, or connection to American car culture, they didn't reach for exotic Italian supercars. They chose Chevrolets with Small Block power because these machines projected the right mix of accessibility and menace. These weren't untouchable fantasies on wheels; they were believable machines that a determined kid could actually own, or build with enough dedication and a weekend job.

George Lucas understood this instinctively when he crafted "American Graffiti" in 1973. The film's Small Block–powered Chevys became visual anchors for the entire narrative, representing youth, freedom, and small-town identity in ways that dialogue alone could never convey. When John Milner's chopped and channeled '32 Ford coupe rumbled through Modesto's neon-lit streets, that sound, the distinctive lope of a performance-cammed Small Block breathing through open exhaust, told audiences everything they needed to know about the character's priorities and self-image. The car wasn't just transportation. It was identity made manifest in steel and sound.

This pattern repeated throughout cinema history. The "Fast and Furious" franchise, beginning decades after "American Graffiti," leveraged vintage Camaros, Novas, and Chevelles to create deliberate contrast with high-tech imports and modern exotics. When Dom Toretto stood beside a black 1970 Chevelle, the filmmakers were making a statement about roots, loyalty, and old-school power. That car, and its Small Block V8, represented a philosophy about how power should feel: raw, mechanical, visceral. Not refined through computers and sensors, but delivered through carburetors, mechanical lifters, and the driver's right foot.

The History of the Small Block Chevy Motor

Sound design teams understood the Small Block's cinematic value perhaps better than anyone. In chase scenes and drag race sequences, audio engineers leaned heavily on the distinctive tonal character of a tuned Small Block, especially through open exhaust. That sound, part mechanical aggression, part controlled chaos, became shorthand for tension, anticipation, and impending action. Listen closely to 1970s and 1980s chase-heavy TV movies and B-films, and you'll hear the same audio signatures repeated: the bark of throttle openings, the distinctive resonance of dual exhausts, the mechanical clatter of solid lifters at idle. Directors knew that audiences had learned to associate those specific sounds with excitement.

Stunt coordinators made pragmatic choices that happened to reinforce the Small Block's cultural dominance. When planning elaborate chase sequences, car crashes, or jumps, coordinators repeatedly chose Small Block–based Chevrolets because they could be built strong, maintained easily between takes, and repeatedly abused on set without catastrophic failures. Studios could keep multiple identical cars in rotation, ensuring visual continuity even as the physical punishment mounted. This practical consideration, choosing the Small Block because it was reliable and repairable, inadvertently cemented its place as the engine of action cinema.

When you think of a movie car that really stuck with you, was it a modern supercar, or something like a Chevelle, Nova, or Camaro? What did that vehicle communicate about the character driving it? And here's the deeper question: how much of your memory of those scenes is visual, and how much is tied to the specific sound of the engine? If you're being honest with yourself, you'll probably realize that the audio component, that Small Block rumble, occupies a disproportionate space in your memory compared to the visual details.

Section 9.2: Television's Small Block Heroes

On television, Small Block–powered cars often became recurring characters in their own right, embodying the personality of the show and the values of its protagonists week after week.

Television operates differently from film. Where movies can treat cars as set pieces, spectacular but temporary, television relies on repetition and familiarity. When a series features a recognizable Chevy, whether a Camaro, pickup, or full-size sedan, viewers begin to associate that vehicle with reliability, mischief, toughness, or working-class pride. Over the course of seasons, that relationship deepens. The vehicle becomes as much a part of the show's identity as the human cast.

Consider the recurring TV muscle car, the Camaro or Chevelle, that appears in crime dramas or action series. Its presence typically signals that the hero is about to break the rules, push the limits, or escape danger through skill and nerve rather than high-tech gadgets. The Small Block's mechanical simplicity translated well to television's narrative needs. There's something inherently dramatic about a character who relies on horsepower and driving ability rather than electronic assistance or artificial intelligence. The engine becomes a co-conspirator in the character's rebellion against authority or convention.

But the Small Block's television presence extended far beyond high-performance muscle cars. Chevy pickups and full-size sedans powered by Small Blocks served as the quiet backbone in family dramas, rural shows, and cop series, rarely mentioned but always there. These vehicles communicated different messages: dependability, unpretentiousness, working-class values. When a character climbed into a Chevy C10 pickup with a 350 under the hood, audiences instinctively understood something about that person's identity and values, even if the truck itself never became a plot point.

Behind the scenes, TV car wranglers favored Small Block–based platforms for practical reasons that reinforced their on-screen presence. Because Small Block Chevys were plentiful and easy to maintain, studios could keep multiple identical cars in rotation, ensuring continuity even as stunt driving, jumps, and practical effects took their toll. Parts availability meant that repairs between episodes could be handled quickly and cost-effectively, an important consideration given television's tighter budgets and faster production schedules than feature films.

Over time, certain vehicles transcended their mechanical roles and became emotional anchors for their shows. Fans knew that when "that car" showed up on screen, whether a detective's personal ride or a family's trusted pickup, something important was about to happen. The vehicle's appearance triggered anticipation, the same way a particular musical theme might signal a character's entrance.

Think back to a TV show you watched regularly during your formative years. Was there a vehicle you remember just as clearly as any human character? What did its make, model, and attitude say about the world of that show? If you've ever owned a similar car or truck in real life, how did that connection shape the way you experienced the show? For many enthusiasts, the overlap between fictional vehicles and real-world ownership created a powerful feedback loop: the TV car validated their choices. At the same time, their personal experience with the platform deepened their connection to the on-screen version.

Section 9.3: Music, Myth, and the Small Block Soundtrack

In rock, country, and even some mainstream pop, the Small Block Chevy became shorthand for a specific kind of American dream, one built on open roads, late-night wrenching, and the thrill of acceleration.

Songwriters didn't always name the Small Block directly, but when they sang about Chevys, main street cruising, and building a motor in the driveway, they were almost always invoking Small Block imagery, because that's the engine most listeners imagined under the hood. The specificity didn't matter as much as the shared understanding. When Bruce Springsteen sang about racing in the street or Don Henley referenced a "Boys of Summer" Corvette, the audience supplied the mechanical details from their own cultural knowledge.

In rock and country music, the car often stands in for larger themes: independence, escape, defiance. A built Small Block, with its loping cam and distinctive exhaust note, becomes the mechanical voice of those emotions, loud, imperfect, unmistakably alive. The engine represents unfinished business, personal expression, and the refusal to accept stock specifications as the final word. Building that motor in your garage, swapping that cam, tuning those carbs, these weren't just mechanical tasks. They were acts of self-definition.

Country music, particularly, embraced the Small Block as a symbol of working-class authenticity. Songs referenced building motors, swapping cams, or polishing chrome on an old Chevy pickup, activities that signaled both mechanical competence and a value system that prioritized hands-on skill over abstract credentials. The Small Block became emblematic of a particular kind of American masculinity: practical, capable, rooted in tangible achievement rather than theoretical knowledge.

Album art and stage design reinforced these connections. Band logos pulled visual cues from valve covers and air cleaners. Stage sets featured Camaros, Novas, or Chevelles, often with engines prominently displayed or revved as part of the performance itself. Some artists recorded real engine sounds, frequently Small Block Chevys, for use as intros, interludes, or background textures on albums and live shows. That audio signature became as recognizable as a guitar riff, instantly communicating mood and attitude.

When you hear a song about "my old Chevy," what thought forms in your mind? Is it a stock sedan fresh from the dealership, or a lightly rough-looking, small-block-powered car with headers and glasspacks? How much of your emotional response to that kind of music is tied not just to lyrics, but to the implied sound and feel of the car being described? These questions matter because they reveal how deeply the Small Block's character has penetrated our shared cultural imagination.

Section 9.4: The Written Word: Literature and Automotive Journalism

Authors, magazine editors, and automotive journalists helped codify the Small Block's myth, turning individual stories of hot rods, street races, and family road trips into a shared narrative of what this engine represents.

Throughout the 1960s, '70s, and beyond, enthusiast magazines regularly featured Small Block builds, drag strip reports, and technical how-tos. Publications like Hot Rod, Car Craft, and Popular Hot Rodding created what amounted to an ongoing narrative about the Small Block's capabilities and cultural significance. Step-by-step 350 rebuild series made engine building feel achievable for a motivated home mechanic. Road tests of new Camaros, Corvettes, and Chevelles praised "small-block punch" and throttle response, creating a consistent vocabulary around the engine's character.

Over time, these articles created a sense that the Small Block wasn't just an engine; it was the default platform for anyone serious about going faster. Alternatives existed, certainly, but they required justification. The Small Block needed no explanation; it was the baseline against which everything else was measured. This framing had enormous influence on how enthusiasts approached their own projects. When you started with a Small Block, you were working within a known framework, supported by decades of accumulated knowledge and a parts infrastructure designed specifically around your needs.

Non-fiction books and memoirs often use a Small Block–powered car or truck as the backdrop for coming-of-age stories, father–child bonding, or the shift from naive enthusiasm to hard-earned mechanical skill. The engine serves as both technical challenge and emotional anchor. Learning to rebuild that 350 becomes a metaphor for broader lessons about patience, attention to detail, and the satisfaction of completing difficult work with your own hands.

Writers leaned heavily on specific technical details to convey authenticity: casting numbers, carburetor sizes, camshaft specifications. These details, repeated over decades across thousands of articles and books, helped educate readers while further embedding the Small Block into enthusiast identity. When you encountered phrases like "bulletproof 350," "trusty small block," or "Chevy V8 under the hood," you weren't just reading a technical description; you were absorbing shared cultural knowledge about what those configurations represented and what they made possible.

The journalistic language around the Small Block created expectations and shaped aspirations. When magazines described a particular combination as "streetable" or praised an engine's "broad torque curve," they were teaching readers how to evaluate performance in practical terms, not just theoretical peak numbers. This emphasis on real-world usability, a reflection of the Small Block's own design philosophy, influenced generations of builders and modified countless project cars.

Have you ever read a car feature or memoir where the engine felt like a co-star in the story? What specific details made it come alive on the page? When you see the phrase "built small block" in print, what assumptions do you immediately make about the car and the person who built it? Your answers probably include assumptions about hands-on involvement, practical goals, and a certain mechanical philosophy, all elements of the Small Block's cultural identity that were shaped and reinforced through decades of automotive writing.

Section 9.5: Video Games, Simulators, and the Digital Small Block

Racing games and driving simulators gave the Small Block a second life with audiences who might never have driven a carbureted

V8 in the real world, preserving its image and sound in a fully digital medium.

As video games evolved from simple sprites to realistic 3D environments during the 1990s and 2000s, developers began licensing real cars and recording real engines. Classic Camaros, Novas, Chevelles, Corvettes, and pickups, many originally powered by Small Blocks, became selectable vehicles in mainstream racing titles. For younger players, this was often their first exposure to the feel of an old-school American V8, even if it was mediated through a controller.

The games translated concepts like torque curves, weight transfer, and traction limits into learning experiences about how these cars behave differently from modern performance vehicles. Players discovered that Small Block–powered muscle cars required different techniques than high-revving imports or sophisticated European sports cars. You couldn't simply pin the throttle and expect traction; you had to modulate power delivery, anticipate weight shift, and respect the limitations of 1960s and '70s chassis technology. These lessons, learned in virtual environments, often created an appreciation that later translated into real-world enthusiasm.

Console and PC racing franchises regularly included classic Chevys in their car lists, typically highlighted in categories like "muscle," "vintage," or "street racing." Upgrade trees allowed players to mimic real-world Small Block modifications: camshafts, cylinder heads, intake manifolds, exhaust systems, and forced induction. This digital version of hot rodding preserved the Small Block's core appeal, incremental improvement through informed choices, while making it accessible to people who lacked the garage space, tools, or budget for the real thing.

Modding communities took this even further. Dedicated enthusiasts created custom vehicles and engine swaps inside digital platforms, echoing the real-world "350 swap" phenomenon and transferring hot-rodding culture into virtual garages. Modders approximated torque curves, recorded and implemented authentic sound samples, and shared their creations with global communities. This activity represented more than just gaming; it was cultural preservation and transmission happening in the digital space.

The educational side effects shouldn't be underestimated. Players developed practical understanding of why the Small Block earned its reputation by experiencing the relative differences between naturally aspirated torque and high-revving import characteristics, between pushrod simplicity and overhead-cam complexity. They learned, through direct digital experience, that the Small Block's appeal wasn't just about peak horsepower numbers, but about the delivery, the character, the way power arrived and was sustained across the rev range.

If your first "drive" in a classic Chevy happened in a video game, how did that shape your expectations when you eventually rode in or drove a real one? What does it say about the Small Block's cultural status that game developers considered it essential to include these cars, even decades after their original production peak? The answers reveal something important: the Small Block's significance transcends pure technical merit. It occupies a place in cultural memory that newer, more powerful engines haven't matched, despite their objective advantages.

Section 9.6: Collectibility, Museums, and Cultural Preservation

By the time museums, auction houses, and major collections began formally recognizing the Small Block's importance, they were acknowledging more than technical achievement; they were preserving a shared cultural memory.

Small Block–powered vehicles that played key roles in films, racing history, or pivotal cultural moments often command significant premiums at auction, even when technically similar cars remain relatively affordable. A 1970 Chevelle with documented film provenance might sell for multiples of what a mechanically identical car would bring, simply because it carries cultural significance beyond its mechanical specification. Collectors aren't just buying transportation or even performance; they're acquiring a tangible link to shared memories and cultural moments.

Museums routinely highlight both the engineering story and the broader cultural context when displaying Small Block–powered vehicles. Interpretive panels explain the lightweight design, favorable power-to-weight ratios, and durability that made the engine successful. But they also discuss car clubs, street scenes, movies, and music of the era, recognizing that the Small Block's importance can't be separated from the cultural movements it enabled and embodied.

Individual engines, especially those with documented racing or film history, are sometimes preserved on stands and treated almost like sculptures or historical artifacts rather than just worn-out mechanical assemblies. Early 265 and 283 engines, Corvette powerplants from significant model years, and legendary race motors occupy museum space alongside the complete vehicles they once powered. This preservation recognizes that the engines themselves, independent of the cars they inhabited, represent important moments in automotive and cultural history.

Private collectors often specialize in Small Block–era Chevrolets, curating cars to represent specific cultural snapshots: cruising culture, drag racing, or the experience of walking into a Chevrolet dealership in 1969 and seeing performance options lined up on the floor. These collections preserve not just vehicles, but the possibility space those vehicles represented, the sense that with the right choices and enough determination, you could build something exceptional.

Auction results consistently demonstrate that cultural significance rivals technical specification in determining value. A numbers-matching, period-correct Small Block–powered car with documented history commands premiums that reflect more than scarcity or performance capability. Buyers are paying for authenticity, for connection to a particular moment in time, for the ability to experience, or re-experience, what those cars represented when they were new or at their cultural peak.

The History of the Small Block Chevy Motor

When you see a Small Block–powered car behind ropes in a museum, what exactly do you feel you're looking at: a piece of engineering, or a physical link to a time, place, and way of life? How do you personally weigh cultural significance versus pure performance when you think about which cars deserve preservation? These aren't abstract questions. They shape preservation priorities, auction values, and which aspects of automotive history get remembered and transmitted to future generations.

In this chapter, you've traced how the Small Block Chevy stepped out of the engine bay and onto the cultural stage. You've seen how films and television used Small Block–powered cars as visual and sonic symbols of attitude, era, and identity, from "American Graffiti's" street cruisers to the muscle cars that defined action cinema for decades. You've discovered how music, literature, and journalism wove the Small Block into stories of freedom, rebellion, and coming-of-age, creating a shared vocabulary that transcended technical specification.

You've explored how video games and simulators introduced the engine's character to new generations in digital form, preserving its distinctive appeal for players who may never turn a wrench on a physical V8. And you've examined how museums, collectors, and enthusiasts now preserve these engines and cars as artifacts of cultural as well as mechanical history.

The chapter's guiding idea was simple but powerful: truly successful engineering doesn't just solve a technical problem, it reshapes how people imagine possibility. The Small Block Chevy began as a compact, efficient V8 designed for mass-market transportation. Over time, it became a symbol: of performance you could build in your own garage, of an America defined by the open road, and of a mechanical soundtrack that still stirs emotion decades later. That transformation didn't happen through marketing alone. It emerged from the engine's presence at countless cultural moments, captured on film, celebrated in song, preserved in print, and now immortalized in digital form.

The History of the Small Block Chevy Motor

The connections run deep. The racing dominance you explored in Chapter 6 fed Hollywood's love affair with Small Block–powered cars; directors knew these were winning engines, not just attractive props. The custom culture documented in Chapter 7 provided the visual vocabulary that musicians and filmmakers repeatedly referenced; those modified Chevys weren't just in movies and songs, they were parked in driveways across America. The global reach examined in Chapter 8 meant that the Small Block's cultural influence extended beyond U.S. borders, appearing in international films, music, and media as a symbol of American automotive character.

Next, you'll step back and look at the entire arc of the Small Block story, from its 1950s origins to the LS and Gen V eras, and into a future increasingly shaped by electrification. Chapter 10 will connect the cultural legacy you've just explored with the engine's ongoing technical evolution, asking a crucial question: in a changing automotive landscape, what does it really mean for the Small Block to endure?

When you hear that familiar V8 rumble, on a movie soundtrack, in a song, or echoing down a real street, you're not just hearing combustion. You're hearing decades of stories, dreams, and late-night work under dim shop lights. The Small Block Chevy may be made of iron, aluminum, and oil. Still, in popular culture, it runs on something less tangible and far more durable: the collective imagination of everyone who ever believed that the right car, with the right engine, could change their life.

The History of the Small Block Chevy Motor

Chapter 10: The Enduring Legacy and Future

Think about two garages, sixty years apart. In 1965, a teenager leans over a grease-stained fender, tuning the carburetor on a 327, adjusting the mixture screws by ear, feeling the engine's rhythm through the air's vibration. In 2025, another young enthusiast stands at a workbench, unboxing a brand-new LS crate engine, its aluminum castings gleaming under LED shop lights, a laptop nearby loaded with tuning software. Different decades, different tools, but the same unmistakable heartbeat is about to come to life.

After tracing the Small Block's journey from factory floors to racetracks, from movie screens to car shows, we now step back and ask a larger question: what does all of this add up to, and where does the story go from here? The Small Block Chevy isn't just automotive history; it's living technology, still evolving, still relevant, still shaping how enthusiasts think about performance and possibility.

This chapter examines the Small Block's enduring influence through multiple lenses: its unprecedented production legacy, the modern LS and Gen V continuation, factory and aftermarket support systems, engineering principles that transcend generations, environmental adaptation, the emerging electric future, preservation efforts, and the enthusiast community that keeps the heartbeat strong. By chapter's end, you'll understand not just what the Small Block has meant, but what role it might play in your own automotive future, and how you can contribute to a legacy that spans seven decades and counting.

Section 10.1: A Production Legacy Without Equal

When we talk about significant internal combustion engines, truly transformative powerplants that shaped automotive history, we're usually discussing a handful of designs: the Ford flathead V8, perhaps the Chrysler Hemi, maybe the Volkswagen flat-four. But when we talk about production volume, sustained manufacturing, and sheer industrial impact, the Small Block Chevy occupies a category almost by itself.

The numbers tell a story that's hard to grasp fully. From 1955 through 2003, Generation I Small Blocks alone accounted for more than 90 million engines across all displacement variants, applications, and markets. That's not a typo, ninety million. Add in the LS family from 1997 forward, and the Gen V engines still in production, and conservative estimates place total Small Block family production well past 100 million units. To put that in perspective, that's roughly equivalent to producing one engine for every three people currently living in the United States.

But production volume isn't just a matter of bragging rights or record books. Those tens of millions of engines created something more valuable than any single powerplant could: an ecosystem. Parts availability sustained over decades means that whether you're working on a 1955 265 or a 2020 LT1, you can source components, often from multiple suppliers, at competitive prices with next-day delivery. That kind of support infrastructure simply doesn't exist for most engines, even successful ones.

The knowledge base matters just as much as the parts network. When millions of mechanics, racers, and hobbyists have worked on essentially the same architecture, when the fundamental rebuild procedures, bearing specifications, and clearance requirements remain recognizable across generations, you're never truly working alone. There's always someone who's solved your problem before, documented the solution, and posted it somewhere accessible.

Think about what that production timeline actually represents. The first 265 V8 rolled off Chevrolet's assembly line in late 1954, beginning volume production for the 1955 model year. Within just a few years, Chevrolet hit the one-millionth Small Block milestone, a production pace that would have seemed impossible for a brand-new design. But the engine worked, and more importantly, it kept working. Municipal fleets standardized on Small Block power because when a police car or utility truck needed an engine replacement, the part was available, the mechanics knew the procedure, and the vehicle was back in service quickly. That reliability in fleet service, where downtime means real money lost, proved the design's fundamental soundness in a way no advertising campaign could match.

Generation I production continued, with periodic updates and displacement changes, until 2003 in North American truck applications. Think about that span: nearly fifty years of continuous production for what was essentially the same basic architecture. During that time, competing manufacturers underwent multiple complete engine redesigns, chasing overhead-cam configurations, exotic materials, and complex technologies. The Small Block kept pace with evolving performance and emissions requirements through incremental refinement of the original concept, proving that Ed Cole's team had gotten the fundamentals right from the start.

The transition periods tell their own story. When new generations overlapped with existing production, LS engines appeared alongside traditional Small Blocks in the late 1990s, Gen V variants coexisting with LS powerplants in the 2010s; it wasn't because GM couldn't make up its mind. It reflected the engine's versatility and the reality that different applications demanded different solutions. A work truck didn't need cutting-edge technology; it needed proven durability and easy service. A Corvette demanded maximum performance and efficiency. The Small Block family could serve both needs simultaneously.

Consider the implications for your own project. When you're selecting an engine, that massive production legacy translates directly into confidence. You're not working with orphan technology or gambling on parts availability five years down the road. You're building on a foundation supported by over half a century of continuous development, millions of successful applications, and a parts ecosystem that shows no signs of disappearing. Whether you're doing a straightforward restoration, planning a performance build, or designing a completely custom application, the Small Block's production legacy provides a safety net that few other engine platforms can match.

When you think about an engine's importance, do peak horsepower numbers really tell the whole story, or does widespread, long-term use matter more? The Small Block's production legacy suggests that real significance isn't measured by dynamometer peaks but by decades of reliable service across millions of applications. That's not just industrial achievement, it's earned trust, repeatedly validated by people betting their time, money, and projects on an engine that simply refuses to become obsolete.

Section 10.2: Understanding Your Engine's Place in the Family Tree

Every Small Block has a story, a specific place in that vast production history. Understanding where your particular engine fits, whether it's a numbers-matching original, a replacement unit, or a modern successor, helps you make better decisions about maintenance, modification, and long-term planning.

Start by identifying your engine's basic specifications: displacement, year of manufacture, original application, and any major running changes that affect parts compatibility. A 1968 327 and a 1985 350 share a fundamental architecture, but they differ significantly in details, camshaft profiles, valve sizes, combustion chamber volumes, and accessory mounting. Those differences aren't problems; they're simply facts you need to work with.

The History of the Small Block Chevy Motor

Create what might be called an "engine pedigree" for your project, a one-page summary documenting where your powerplant fits in the larger Small Block timeline. Note the production years for your specific variant, the primary vehicle applications, any known running changes during the production span, and the typical performance characteristics. This reference becomes invaluable when sourcing parts, troubleshooting problems, or planning modifications. You'll know immediately whether a specific component is correct for your application or if you're looking at a part from a different generation.

That knowledge also builds confidence. When you understand that your engine is one of millions sharing the same core architecture, it changes how you approach problems. You're not dealing with mysterious, poorly documented technology. You're working with one example of a thoroughly proven design, and somewhere out there, someone has already figured out the solution to whatever challenge you're facing.

Walk past a modern LS3 or Gen V LT4 on a display stand, and you might not immediately recognize it as a Small Block. The aluminum castings look sculptural and contemporary. Coil-near-plug ignition eliminates the traditional distributor. Composite intake manifolds and tucked-away accessories create clean lines that bear little resemblance to the iron-block, cast-manifold appearance of a classic 350. But look closer, really study the architecture, and the family resemblance becomes unmistakable.

The LS and Gen V engines represent something rare in automotive engineering: a genuine continuation rather than a replacement. When GM could have abandoned pushrod V8s entirely, following much of the industry toward overhead-cam designs, they instead asked a harder question. What if we kept the fundamental Small Block philosophy but executed it with modern materials, manufacturing processes, and electronic management? What if we proved that the original concept was sound enough to meet 21st-century demands?

The History of the Small Block Chevy Motor

The core DNA remains intact. Pushrod actuation with two valves per cylinder delivers simplicity, packaging efficiency, and low reciprocating mass. The compact external dimensions mean the engine fits in spaces that overhead cam designs simply can't match, a crucial advantage for everything from sports cars to trucks. The strong bottom end, with robust main bearing support and a stiff crankshaft, provides a foundation for both high output and long service life.

But the execution leverages everything engineers have learned in half a century of Small Block development. Aluminum blocks and heads, once exotic, costly options, became standard in many LS and Gen V variants, cutting weight dramatically while maintaining structural integrity. Deep-skirt block designs and improved oiling systems addressed weaknesses discovered through decades of racing and fleet service. Coil-near-plug ignition eliminated the distributor's mechanical complexity and provided precise spark control. More rigid bottom ends, cross-bolted mains, and improved windage control supported power levels that would have seemed impossible to Ed Cole's original team.

The engineering targets remained recognizably Small Block: high specific output, exceptional durability, easy manufacturability, and broad adaptability. The LS3, generating 430 horsepower from 376 cubic inches in naturally aspirated, emissions-compliant street trim, delivers better than 1.14 horsepower per cubic inch. This figure puts it in supercar territory by the standards of just two decades earlier. The LS7, with its 427 cubic inches and titanium intake valves, produces 505 horsepower while meeting modern emissions standards and providing factory warranty coverage. These aren't fragile, high-strung race engines; they're production powerplants that start reliably in sub-zero temperatures and cruise effortlessly at modern highway speeds.

Gen V engines pushed the envelope further. Direct injection delivers fuel precisely into the combustion chamber, improving both power and efficiency. Variable valve timing adjusts cam phasing on the fly, optimizing performance across the entire RPM range rather than compromising with a single fixed setting. Active Fuel Management, cylinder deactivation that lets the engine run on four cylinders during light-load cruising, significantly improves fuel economy without driver intervention. Sophisticated knock control with individual cylinder monitoring enables aggressive timing advance while protecting against detonation.

Yet despite all this technological sophistication, Gen V engines maintain the fundamental Small Block packaging concept. They're not physically larger than their predecessors. Accessory mounting remains straightforward. The basic service procedures, changing oil, replacing spark plugs, and adjusting valves, don't require exotic tools or specialized training. A mechanic who grew up working on Generation I Small Blocks can understand a Gen V's architecture, even if the electronic control systems require new knowledge.

Consider a direct comparison: a classic 350 in a 1970 Chevelle beside an LS3 in a modern Camaro or high-end restomod. Both displace similar volumes, 350 versus 376 cubic inches. Both use pushrod valve actuation and two valves per cylinder. Both measure similarly in external dimensions. But the LS3 generates over 400 horsepower in smog-legal street trim, delivers better fuel economy, requires less frequent maintenance, and produces significantly fewer emissions, all while weighing substantially less than its cast-iron predecessor.

Real-world builders who've swapped from carbureted Gen I Small Blocks to LS-based setups report consistent improvements. Weight drops noticeably, aluminum LS engines typically weigh 100-150 pounds less than equivalent iron small-blocks, improving weight distribution and handling response. Drivability transforms, with smooth idle quality, crisp throttle response, and excellent cold-start behavior replacing the compromises inherent in carbureted setups. Fuel economy often improves by 20-30 percent, making long-distance cruising more practical and affordable. And reliability, already a Small Block strength, reaches new levels, with 200,000-mile service lives common in stock form.

Does all this technology mean the LS and Gen V are really "new" designs, completely separate from their predecessors? Or are they exactly what they appear to be: the Small Block philosophy executed with modern tools and knowledge, proof that the fundamental concept was sound enough to evolve rather than being replaced?

Section 10.3: Choosing Your Path: Tradition Versus Technology

The coexistence of multiple Small Block generations creates both opportunity and decision points for enthusiasts. Do you stick with a Gen I or Gen II Small Block for your project, honoring period correctness and mechanical simplicity? Or do you embrace an LS or Gen V swap to gain modern performance and efficiency?

Neither answer is inherently right or wrong; it depends on your specific goals and priorities. A numbers-matching restoration obviously demands the original engine configuration. But for driver-quality classics, restomods, and custom builds, the choice becomes more nuanced.

Based on that honest assessment, not sentiment, not trends, not what you see others doing, which path actually aligns with how you'll use the vehicle? If it's a weekend toy driven 1,000 miles annually and shown at local events where originality matters, the traditional Small Block makes perfect sense. If it's a 10,000-mile-per-year cruiser that needs to keep pace with modern traffic and deliver reasonable fuel economy, the LS or Gen V swap becomes increasingly attractive.

The beauty of the Small Block family tree is that you're not abandoning anything fundamental when you choose modern technology. You're still building on the same core philosophy, just executing it with 70 years of additional knowledge and development. The heartbeat remains recognizably similar, the characteristic V8 burble, the torque curve's shape, the fundamental character that makes a Small Block feel like a Small Block regardless of generation.

The History of the Small Block Chevy Motor

Crate Engines and Factory Support: The Engine That Never Retired

Most engines eventually disappear from factory support pipelines. Production ends, new designs replace old architectures, and within a decade or two, official parts and technical documentation become scarce. The aftermarket may continue to support popular designs, but without factory backing, the knowledge base gradually erodes and component quality becomes inconsistent.

The Small Block followed a different path. Through GM Performance Parts, later rebranded as Chevrolet Performance, the engine effectively remained "in production" for enthusiasts even as original-equipment installations declined. The crate engine program transformed the Small Block from a vehicle component into a modular performance product, available as a complete, warrantied assembly ready for installation in virtually any application.

Crate engines aren't a new concept; hot rodders have been buying complete engines from specialty builders for decades. But factory-supported crate programs operate on a different scale and with different advantages. These are engines assembled on modern production equipment using current manufacturing processes, with the same quality control as OEM installations. They arrive dyno-tested, with documented performance curves, complete installation instructions, and actual factory warranty coverage. You're not gambling on an unknown builder's work or hoping a junkyard core isn't hiding expensive problems; you're buying proven technology with factory backing.

The range is remarkable. At the accessible end, Chevrolet Performance offers straightforward 350 replacement engines, basic, carbureted small-blocks producing around 290 horsepower, perfect for driver-quality classics that need reliable power without complexity. Step up to LS-based crates, and the options multiply: LS3 engines producing 430-480 horsepower depending on configuration, supercharged LSA powerplants generating over 550 horsepower, race-focused variants with forged internals ready for even higher output.

Gen V options extend the range further into modern territory. Direct-injected LT engines are available in crate form, ranging from mild truck engines to 650-horsepower LT4 supercharged variants. These aren't just engines; they're complete powertrain solutions, often available with matching transmissions, controllers, and wiring harnesses, simplifying integration into virtually any chassis.

This factory support fundamentally changes the Small Block's lifecycle. The engine is no longer tied to the production span of any particular vehicle model. A 1950s hot rod can receive a brand-new LS3. A 1960s muscle car can get modern LS or LT power without compromising its character. Kit cars, customs, and even boats gain access to proven, warrantied engines without scrounging junkyards or gambling on rebuilds.

Consider the practical comparison between a crate engine and rebuilding an unknown core. The junkyard Small Block might cost $300-500 initially, but by the time you factor in machine work, boring, honing, decking, valve jobs, new pistons, bearings, gaskets, and the inevitable surprises hidden in that "running when pulled" engine, you're often $2,500-3,500 deep before the first spark plug fires. A basic 350 crate engine runs around $3,500-4,000, delivered to your door, fully assembled, with a warranty and documented specifications. The time savings alone, weeks or months of machine shop backlogs and assembly work, often justify the crate engine cost.

For higher-performance applications, the comparison becomes even more favorable. Building a reliable 450-horsepower Small Block from scratch requires careful parts selection, precise machine work, and experienced assembly. Budget $6,000-8,000 in parts and machine work, plus your time, plus the risk of mistakes or incompatible components. An LS3 crate producing 430 horsepower runs about $7,500- $ 8,500 with aluminum heads, modern fuel injection, and factory testing. You're not paying a huge premium for the convenience; you're buying proven engineering, quality control, and peace of mind.

The existence of comprehensive crate engine programs also legitimizes performance and swap culture in a way that backyard builds never could. When the factory itself offers complete LS swap packages, including engines, transmissions, mounts, harnesses, and installation manuals, it sends a clear message that these projects aren't fringe modifications. They're officially sanctioned uses of the technology, with proper engineering backing and parts support.

Evaluating Crate Versus Rebuild for Your Project

The crate-or-rebuild decision deserves careful, honest analysis rather than assumptions. What matters more to you personally: the process of building an engine yourself from the ground up, understanding every component and clearance? Or having a reliable, dyno-tested foundation you can drop in and drive immediately?

Neither answer is wrong, but they lead to dramatically different project timelines and experiences. If you've never rebuilt an engine and want to learn those skills, building from a core, even if it costs more and takes longer, provides invaluable hands-on education. If you're juggling limited time, want to minimize risk, and prioritize getting on the road, the crate engine offers compelling advantages.

Take a practical step: visit Chevrolet Performance's website or a reputable crate engine supplier. Identify three options that fit your budget range. Compare their horsepower and torque curves, noting where peak numbers occur and what the power bands look like. Consider the recommended applications: some engines optimize for low-end torque in trucks, while others emphasize high-RPM power for racing applications.

Now draft a rough "crate versus rebuild" budget. For the rebuild path, include realistic estimates for machine work,$800-1,200 for a basic freshening, $1,500-2,500 for a complete job with performance upgrades. Add parts, pistons, rings, bearings, gaskets, timing components, realistically running $1,000-2,000 for quality pieces. Factor in your time, dozens of hours for disassembly, cleaning, assembly, and tuning. Note potential wild cards: cracks discovered during magnafluxing, worn components requiring replacement, compatibility issues with mixed-generation parts.

For the crate engine path, the math is simpler but includes often-overlooked elements. The engine itself is the major cost, but you'll also need installation components: motor mounts (especially if you're swapping an LS into an older chassis), headers or exhaust manifolds compatible with your chassis, fuel system components matched to the engine's requirements, and cooling system components. EFI swaps add a complete harness and controller, and may include a new fuel tank or in-tank pump.

Run the numbers honestly. Then, and this is crucial, consider which path you'll actually complete versus which one sounds better in theory. A drawer full of Small Block parts and an engine stand gathering dust in the corner represents more money wasted than a crate engine that has you driving within a few weekends.

Would a crate engine change your timeline or confidence level for a project you've been hesitant to start? Sometimes, the difference between a dream that stays in your head and a car on the road is simply reducing the number of variables, unknowns, and potential failure points. The Small Block's crate engine ecosystem exists precisely to enable more, not fewer, projects to turn "I wish I could" into "I'm going to."

Section 10.4: Engineering Lessons from a Seventy-Year Success Story

The Small Block's longevity isn't an accident of history or the result of corporate stubbornness. It reflects a set of robust engineering principles, modularity, simplicity, adaptability, and manufacturability that modern designers can still learn from, regardless of whether they're creating internal combustion engines, electric drive units, or completely different technologies.

Start with compact packaging. The Small Block's external dimensions remained remarkably consistent across generations precisely because Ed Cole's team understood that engines don't exist in isolation; they must fit within chassis structures, leave room for steering and suspension components, and allow practical access for service and maintenance. By keeping the valvetrain compact through pushrod actuation and designing efficient combustion chambers that didn't require massive cylinder heads, they created an engine that could power everything from compact cars to full-size trucks without requiring dedicated platform architectures.

Structural strength in the block and bottom end provided a foundation for evolution rather than obsolescence. That original 265 design proved capable of growing to 400+ cubic inches through bore and stroke increases, precisely because the core structure was overbuilt relative to the initial power targets. The engineers weren't future-proofing in the modern sense; they were simply designing with appropriate safety margins and sound mechanical principles. But that inherent strength meant the architecture could absorb decades of power increases without fundamental redesign.

The valvetrain architecture balanced cost, capability, and complexity. Overhead valve design with pushrod actuation wasn't the simplest possible approach; flathead designs with side valves are mechanically simpler. Nor was it the most sophisticated; overhead cam designs can support higher RPM limits and more complex valve-timing events. But pushrods offered an optimal middle ground: better breathing than flatheads, sufficient RPM capability for street and moderate racing use, dramatically simpler manufacturing than overhead cam designs, and packaging efficiency that overhead cams couldn't match.

Modularity at the component level enabled incremental improvement without complete redesign. Cylinder heads could be upgraded independently, with better castings, larger valves, and improved port shapes, bolting directly to existing blocks. Intake manifolds, from single-plane racing pieces to dual-plane street designs to fuel-injected variants, all shared common mounting patterns. Accessory drives, ignition systems, and cooling components could be modified or upgraded without altering the engine's fundamental architecture. This modularity meant that racing development, fleet service feedback, and enthusiast modification all fed back into factory revisions and new generations, creating continuous improvement rather than wholesale reinvention.

These principles map directly onto modern engineering challenges, even in technologies that seem completely different from traditional V8s. Electric vehicle platforms increasingly emphasize modularity, scalable battery modules that accommodate different range requirements, motor units that can be configured for rear-drive, front-drive, or all-wheel-drive applications, and skateboard-style chassis structures that support multiple body styles. The specific technology differs fundamentally from the Small Block, but the design philosophy that creates a robust foundation that accommodates variation without complete redesign remains strikingly similar.

The feedback loop principle matters equally across technologies. The Small Block improved continuously because information from diverse applications, NASCAR racing, drag strips, cross-country fleet service, and backyard modifications fed back to factory engineers who refined subsequent versions. Modern development processes formalize this concept through data logging, over-the-air updates, and beta testing programs. Still, the underlying principle remains constant: designs improve when developers receive real-world feedback from varied use cases and incorporate those lessons into ongoing development.

Consider a practical example of the Small Block's modular evolution. The initial 265 design featured specific bore spacing, the center-to-center distance between cylinders. That dimension remained fundamentally constant even as displacement increased, meaning many components remained interchangeable across variants. Pistons changed, bore sizes increased, but the basic block deck dimensions stayed recognizable. This consistency enabled the massive parts ecosystem and knowledge base that makes Small

Block work accessible; you're not learning a completely new engine every time displacement or configuration changes.

A modern parallel exists in automotive electronics platforms. Manufacturers increasingly build scalable electrical architectures, common wiring harnesses, shared control modules, and consistent communication protocols that accommodate everything from base-model economy cars to high-performance variants. The specific sensors and actuators might differ, but the underlying structure remains consistent, reducing development costs, simplifying service, and enabling rapid deployment of new features without redesigning the entire system.

Section 10.5: Applying Modular Thinking to Your Own Project

The Small Block's engineering lessons translate directly into practical project planning. Whether you're building an engine, designing a complete vehicle, or tackling any complex mechanical system, applying modular thinking from the outset makes future changes easier and reduces the risk of boxing yourself into corners.

Take your current engine or project plan and identify three areas where a modular or flexible approach would make future changes easier. The fuel system provides an obvious example: a return-style EFI setup with an adjustable regulator can accommodate power levels from mild to wild by simply changing injector size and pump capacity. Install a restrictive returnless system optimized for stock power, and you'll need a complete replacement when performance goals increase.

Ignition systems offer similar choices. A traditional distributor is simple and familiar, but switching from points to electronic ignition, or from vacuum advance to locked timing for boost, requires a different distributor or extensive modifications. A modern coil-near-plug or distributor-less ignition controlled by programmable EFI eliminates those constraints; timing curves are just software, endlessly adjustable without changing hardware.

Exhaust systems, wiring harnesses, cooling system capacity, transmission selection, every major system includes decision points where you can choose locked-in simplicity or adaptable flexibility. Neither approach is universally correct, but understanding the trade-offs allows you to make informed choices rather than discovering limitations after substantial investment.

List at least one specific change you could make now to avoid boxing yourself in later. That might mean running a wiring harness with extra circuits you don't currently need, leaving space in the engine bay for a future turbocharger even if you're starting naturally aspirated, or choosing a fuel system rated for 50 percent more power than your current combination produces. These decisions often cost little additional money upfront, maybe 10-15 percent more than the bare minimum, but they preserve options that become expensive or impossible to retrofit later.

Think about complex systems you've worked on where you benefited from someone else's foresight, or struggled because someone chose the most restrictive possible approach. The Small Block's seventy-year success story demonstrates that designing for adaptability, even when the specific future applications remain unknown, creates technology that outlasts its original context and continues serving needs its designers never imagined.

If you were designing an engine or any complex system today, what aspects would you intentionally overbuild or standardize to allow for future evolution? The Small Block's lesson isn't that you need to predict the future perfectly. It's those robust fundamentals, thoughtful modularity, and realistic safety margins that create technology that can evolve rather than become obsolete.

Emissions, Efficiency, and Environmental Pressures

The Small Block's continued survival into the emissions era, and then into the high-efficiency, electronically managed era, demonstrates something crucial about engineering: fundamentally sound mechanical designs can adapt to regulatory and environmental demands without losing their core identity. This wasn't guaranteed or easy, and the adaptation process substantially reshaped the engine. But the fact that recognizably Small Block powerplants still pass modern emissions testing while delivering impressive performance proves the original architecture's inherent flexibility.

The regulatory arc began in earnest during the 1970s, as federal and California emissions standards started constraining what had been a virtually unregulated industry. Early responses were crude: lean carburetor calibrations, reduced compression ratios, and massive air pumps forcing fresh air into exhaust manifolds to burn unburned hydrocarbons. Performance suffered dramatically. A smog-era 305 producing 145 horsepower from 305 cubic inches represented embarrassing specific output compared to pre-emissions Small Blocks that routinely generated over 300 horsepower from similar displacement.

But engineers learned. Catalytic converters, initially crude, restrictive devices, evolved into sophisticated three-way units that simultaneously reduced oxides of nitrogen, carbon monoxide, and unburned hydrocarbons. Electronic fuel injection replaced carburetors, allowing precise fuel metering that improved both emissions and drivability. Oxygen sensors enabled closed-loop feedback, letting the engine continuously adjust fuel delivery to maintain ideal combustion. Onboard diagnostics monitored system operation, detecting problems before they caused emissions failures or drivability issues.

Each regulatory layer added complexity, but thoughtful engineering transformed those constraints into opportunities. Improved combustion chamber designs, heart-shaped chambers that promoted faster, more complete burning, reduced emissions, while recovering lost power. Roller camshafts reduced friction and enabled more aggressive lobe profiles, improving efficiency. Better sealing through improved gaskets and closer machining tolerances reduced oil consumption. Tighter bearing clearances and improved oiling systems enhanced durability while reducing parasitic losses.

Gen V engines pushed adaptation further still. Direct injection delivers fuel at 2,000+ PSI directly into the combustion chamber during the intake and compression strokes, creating more time for fuel atomization and better mixture preparation. This allows higher compression ratios, previously limited by knock, while reducing particulate emissions. Variable valve timing adjusts cam phasing dynamically based on operating conditions, optimizing valve events for cold start, cruising efficiency, or maximum power without compromise. Cylinder deactivation, Active Fuel Management in GM's terminology, shuts down half the engine during light-load cruising, significantly improving highway fuel economy without driver intervention or performance sacrifices when all cylinders reactivate.

The trade-offs engineers manage are real and sometimes painful. Power versus emissions, cost versus complexity, heritage versus innovation, these aren't abstract debates but concrete engineering challenges with significant consequences. But the Small Block's evolution demonstrates that trade-offs aren't always zero-sum. A modern Gen V LT4, producing 650 supercharged horsepower while meeting current emissions standards and delivering reasonable fuel economy in normal driving, would have seemed impossible by 1970s standards, yet it exists in production form with a factory warranty.

For enthusiasts, these developments create both challenges and opportunities. Legal and practical considerations for swaps and modifications vary dramatically by region. California's stringent regulations require extensive documentation and the use of approved components for any engine swap, effectively limiting options to CARB-certified combinations. Other states maintain more relaxed approaches, focusing on the vehicle's original standards or applying basic safety and emissions checks without detailed component verification.

Understanding your local regulatory environment isn't optional; it's foundational to project planning. A non-compliant vehicle may be impossible to register, undriveable on certain roads, or subject to significant fines. But within those constraints, opportunities exist to build engines that are both powerful and relatively clean when modern components are used correctly.

A legally compliant LS swap in a late-model chassis, using factory catalytic converters, oxygen sensors, and OEM-style engine management, often passes emissions testing more easily than the original engine it replaces, while delivering better performance and efficiency dramatically. The key is understanding that "performance" and "emissions compliance" aren't necessarily opposing goals. Modern technology creates space for both, if you're willing to embrace components and approaches that earlier generations of hot rodders might have dismissed as unnecessary complexity.

Compare a smog-era 305 producing 145 horsepower with 9.5:1 compression through severely restricted breathing, versus a modern Gen V 6.2L producing 460 horsepower with 11.5:1 compression and direct injection. Similar basic architecture, pushrod V8, two valves per cylinder, similar external dimensions, but radically different specific output, emissions, and fuel economy. The modern engine isn't just more powerful; it's cleaner, more efficient, and more durable. That's not marketing spin, it's measured reality.

Building Clean, Powerful Small Blocks in the Modern Era

Balancing the emotional appeal of a traditional V8 with environmental and regulatory realities doesn't require abandoning performance or character. It does require an honest assessment of what you're building and why, followed by informed choices about technology and components.

Start by identifying the emissions and inspection rules that actually apply where you live. Don't rely on forum rumors or what you "heard somewhere"; contact your state's motor vehicle department or inspection authority directly. Understand how regulations affect engine swaps, modifications, and classic vehicles specifically. Some jurisdictions exempt vehicles past a certain age from emissions testing entirely. Others apply standards based on the engine's year rather than the chassis. Still others require that any swap maintain or improve upon the original vehicle's emissions performance.

With that regulatory framework clear, you can make informed decisions rather than gambling. If you're exempt from testing, your choices broaden significantly, though responsible performance building considers emissions impact regardless of legal requirements. If testing applies, understanding the specific standards guides component selection toward combinations that perform well while remaining compliant.

Create a short list of "responsible performance" upgrades for your specific platform. Modern catalytic converters, designed for higher flow and greater durability than 1980s units, reduce restrictions while effectively cleaning exhaust gases. Wideband oxygen sensor tuning allows precise air-fuel ratio monitoring and adjustment, improving both performance and emissions. Modern ignition systems with individual coil packs or distributorless setups provide a stronger, more consistent spark, enabling leaner mixtures that burn more completely. EFI conversions, even throttle-body systems that maintain a carburetor's simplicity, improve cold starting, drivability, and fuel-delivery precision.

None of these upgrades requires abandoning the Small Block's fundamental character. A properly tuned LS-swapped classic still sounds, feels, and drives like a V8-powered performance car; it just happens to idle smoothly, start reliably in cold weather, return decent fuel economy, and pass emissions testing. Those aren't compromises; they're improvements that make the vehicle more enjoyable to drive and easier to live with over the long term.

Are there areas in your current or planned build where adopting newer, cleaner technology would actually enhance your enjoyment rather than diminishing it? An EFI system that eliminates hot-start headaches and altitude-related tuning issues. Fuel injection that provides instant throttle response and eliminates carburetor flooding. Electronic ignition that starts immediately without cranking, setting the choke, or pumping the pedal. Modern cooling system components that regulate temperature precisely without constant gauge monitoring.

Technology can serve the experience rather than detract from it if you're thoughtful about which innovations actually solve problems you experience versus which ones just add complexity for complexity's sake. The Small Block's adaptation to emissions and efficiency standards demonstrates that performance, reliability, and environmental responsibility aren't mutually exclusive when engineering is done thoughtfully.

Section 10.6: The Electric Future and the Place of the V8

The automotive industry is transforming electric vehicles, from affordable commuter cars to exotic supercars, and is increasingly populating showrooms and streets. Hybrid powertrains proliferate even in traditionally gas-powered segments. Major manufacturers announce target dates for full electrification, with some committing to phase-outs of internal combustion engines within the next decade or two. In this rapidly evolving landscape, where do the Small Block Chevy and traditional V8s generally fit?

The honest answer is complex, nuanced, and dependent on timeframes and specific applications. V8s are shifting from mainstream powerplants to specialty, enthusiast-focused technology. Their role is changing, but their relevance isn't disappearing overnight, or even in the near term.

Current trends are undeniable. EV and hybrid penetration in new vehicle sales grows steadily, driven by improving technology, expanding charging infrastructure, regulatory pressure, and shifting consumer preferences. V8s increasingly concentrate in trucks, high-performance models, and specialty applications where their specific advantages, torque characteristics, towing capacity, extended range without recharging, and established infrastructure remain compelling.

But let's separate hype from reality. Existing Small-Block-Powered vehicles will remain on the roads for decades. A well-maintained LS engine can easily deliver 200,000+ miles of reliable service, representing twenty years or more of typical use. Classic and collectible vehicles, many powered by Small Blocks, operate under different regulatory frameworks than new production vehicles, with most jurisdictions exempting vehicles past certain ages from current emissions and safety standards. The installed base of Small Block vehicles isn't disappearing in five or even twenty years.

Parts and crate engine support will continue as long as meaningful demand exists. Chevrolet Performance's business model doesn't require millions of annual sales to remain viable; it only requires sufficient enthusiast demand to justify continued production. As long as people keep building, restoring, and racing Small Block-powered vehicles, the parts ecosystem will persist. It might shift increasingly toward specialists and enthusiast suppliers as mass-market applications decline, but that's a transition, not an extinction.

Regulatory environments may become stricter, particularly regarding new vehicle production and potentially around low-emission zones in dense urban areas. But most jurisdictions recognize the difference between regulating new vehicle manufacturing and retroactively restricting existing vehicles. Outright bans on internal combustion engines for private passenger vehicles remain unlikely in most regions for the foreseeable future. However, certain use cases, such as driving in the downtown cores of major cities, may face increasing restrictions.

The emotional and experiential factors matter here more than some discussions acknowledge. Sound, vibration, and mechanical character aren't trivial details to many enthusiasts. They're fundamental to the driving experience. An electric vehicle with identical or superior acceleration figures doesn't necessarily provide equivalent subjective satisfaction. That's not irrational or anti-progress; it's recognition that human experience involves sensory and emotional dimensions beyond objective performance metrics.

EVs and V8s will likely coexist, serving different roles for different contexts. A household might own an electric daily driver, quiet, efficient, and convenient for routine transportation, and a classic Small Block-powered weekend car that provides a visceral, engaging driving experience on enthusiast outings. Different vehicles, different purposes, both valued. That's not a contradiction; it's an appropriate matching of tools to tasks.

The path forward for Small Block enthusiasts involves realistic assessment rather than alarmism. The Small Block era is evolving into a more focused, enthusiast-centered chapter rather than ending abruptly. Production V8s in new vehicles may become increasingly rare and specialized. But the existing millions of Small Blocks, the supporting parts ecosystem, and the community of people who value this technology won't vanish simply because new vehicle production shifts toward other powertrains.

Consider a practical example: European cities with established low-emission zones. Enthusiasts in those regions have adapted in various ways, registering classics under historic-vehicle programs with special allowances, restricting use to weekend drives outside restricted zones, or focusing on cleaner builds that meet applicable standards. The regulations created constraints, but didn't eliminate the hobby. People who genuinely value these vehicles find ways to continue enjoying them within whatever framework exists.

Section 10.7: Planning Your Long-Term Relationship with Small Block Power

If the future of internal combustion involves a transition rather than immediate extinction, what does that mean for your personal relationship with Small Blocks? The question isn't rhetorical; it deserves thoughtful consideration specific to your circumstances and timeline.

How do you see your own relationship with internal combustion changing in the next 10-20 years? Will Small Block vehicles remain your primary transportation, or transition to weekend-only hobby use? Does your geographic location seem likely to impose increasing restrictions on internal combustion use, or does it maintain more

The History of the Small Block Chevy Motor

relaxed approaches? What's your realistic timeframe? Are you planning projects you'll complete in the next few years, or thinking about twenty-year plans?

If you plan to keep or build a Small Block-powered vehicle, what role do you imagine it playing in your life: daily driver, weekend toy, track car, or rolling heirloom? Each answer implies different priorities and different optimal approaches. A daily driver demands reliability, efficiency, and compliance with local regulations, pushing toward modern LS or Gen V powertrains with full emissions equipment. A weekend toy operated 1,000 miles annually can embrace more traditional configurations without practical concerns about fuel economy or daily drivability. A dedicated track car operates in an entirely different context, with regulations and priorities specific to motorsport competition.

This planning isn't pessimism or acceptance of defeat; it's realistic stewardship. The Small Block enthusiasts who'll still be driving, building, and enjoying these engines two decades from now will be the ones who adapted thoughtfully to changing contexts rather than denying that change exists. The technology is robust, the parts support is substantial, and the community is passionate. Those foundations support a long future for Small Blocks in enthusiast hands, even as their mainstream automotive role evolves.

Preserving Hardware and Knowledge

The Small Block's future depends on more than surviving engines and available parts. It requires preserving skills, documentation, and hands-on knowledge, an intergenerational transfer that turns mechanical objects into living heritage rather than museum artifacts.

What needs preserving breaks down into several interconnected categories. Physical engines and components form the obvious foundation, actual cores, original parts, and rare variants that document the design's evolution. An early 265 prototype or a factory Z28 302 represents irreplaceable historical information. But preservation extends beyond just "numbers-matching" showcase engines. Even common 350s and LS variants deserve thoughtful storage and maintenance for future use, whether in restorations, swaps, or as teaching tools.

Service procedures, tuning techniques, and troubleshooting experience form equally important but less tangible knowledge that exists primarily in the minds and hands of experienced builders. These skills aren't fully captured in manuals or videos. They're the subtle feel of proper ring gap, the sound of correct valve lash, the ability to diagnose problems through observation and experience rather than just throwing parts at symptoms. This embodied knowledge transfers imperfectly through written documentation; it requires hands-on demonstration and practice.

Factory manuals, period literature, and racing development histories provide crucial context and technical information. Original documentation preserves specifications, procedures, and engineering rationale that may not survive in other forms. Racing and development histories capture why certain changes were made, what problems various approaches solved, and how the engine evolved in response to real-world use. This information isn't just nostalgic; it's practical knowledge that helps current builders avoid repeating past mistakes and understand the reasoning behind various design choices.

Multiple entities play roles in preservation efforts. Museums and institutional collections, like the GM Heritage Center, maintain significant examples of Small Block history, from early prototypes to race-winning engines. These collections preserve physical artifacts and documentation that might otherwise disappear, making them accessible to researchers and enthusiasts.

Private restorers and specialist shops function as distributed preservation networks. The high-end restoration specialist who maintains equipment and knowledge for rebuilding 1960s Z28 engines preserves capabilities that might otherwise vanish. The engine builder who still understands how to rebuild mechanical fuel injection properly preserves specialized knowledge that newer generations might never encounter.

Online archives, forums, and video documentation extend preservation into digital spaces. Well-curated forums capture decades of troubleshooting discussions, build documentation, and technical debates. Video content, when thoughtfully produced, preserves a visual demonstration of procedures that text descriptions struggle to convey. These resources democratize access to information that once existed only in specific geographic locations or specialized shops.

But technology and institutions only partly address preservation needs. The human factor remains crucial. Experienced builders mentoring younger enthusiasts is the most effective mechanism for knowledge transfer. When a seasoned machinist walks an apprentice through their first Small Block rebuild, explaining why certain clearances matter, demonstrating how to torque main caps properly, and showing how to recognize wear patterns, that direct transmission of skill and judgment proves far more effective than any manual or video.

Formal training programs, whether through community colleges, trade schools, or manufacturers, provide structured learning paths. But informal "garage apprenticeships" often work just as well, or better, in enthusiast contexts. The teenager, spending weekends helping an experienced builder and gradually taking on more responsibility and complex tasks, gains knowledge that no classroom alone could provide.

Consider a specific example: a museum restoration of an important Small Block-powered race car. The process obviously preserves the physical vehicle. But properly executed, it also documents the restoration process itself, recording techniques, sources for rare components, challenges overcome, and decisions made. That documentation helps future restorers facing similar challenges. Interviews with original crew members or drivers capture first-hand knowledge that exists nowhere else. The restoration becomes not just the preservation of an object, but the preservation of the knowledge and context that make the object meaningful.

Another case study: a seasoned engine builder taking on an apprentice for a complete Small Block rebuild. The apprentice learns more than just the sequence of assembly steps. They learn why certain checks and clearances matter, not just "this is the spec," but "here's what happens if this is wrong." They observe problem-solving in real-time when something doesn't fit quite right or when measurements fall outside expected ranges. They develop mechanical intuition that comes only from hands-on experience, guided by someone who's made mistakes and learned from them.

Your Role in Preservation

Who first taught you how an engine works, or who would you like to learn from next? That question isn't abstract nostalgia; it's an invitation to recognize your place in preservation networks, whether as a learner seeking knowledge or as someone with experience to share.

What Small Block-related knowledge, stories, or parts do you already have that might be worth preserving or passing on? Maybe it's a collection of period service manuals gathering dust in a closet, information that someone else could use. Perhaps it's specialized experience with a particular engine variant that few others have worked on extensively. Or it could simply be a well-documented build that helps others understand what works and what doesn't.

Choose one preservation action you can take in the next month. Scan and back up old service manuals, even common ones, then upload them to archives or share them with relevant online communities. Physical manuals deteriorate, get thrown out, or disappear; digital preservation ensures multiple copies exist across distributed locations. Record a video or photo walkthrough of your current engine build, documenting not just the glamorous parts but the mundane details, problems encountered, and solutions

The History of the Small Block Chevy Motor

discovered. That real-world documentation proves far more valuable than idealized builds that hide all the challenges.

Or consider human preservation: offer to show a younger enthusiast how to perform basic Small Block tasks. Setting valve lash, changing spark plugs, and checking timing are fundamental procedures that might seem trivially simple if you've done them hundreds of times. But to someone starting with no baseline experience, even basic procedures can seem intimidatingly complex. Your time and willingness to demonstrate make the difference between someone developing confidence and someone giving up in frustration.

Preservation isn't just about safeguarding the past; it's about ensuring the future. Every Small Block enthusiast actively engaged today, learning, building, documenting, and teaching, extends the engine's practical lifespan and cultural relevance. The hardware remains valuable only as long as there are people who understand how to use it, maintain it, and appreciate it. That's not automatic; it requires conscious effort and intergenerational transfer.

Section 10.8: The Enthusiast Community: Keeping the Heartbeat Alive

The Small Block's legacy is sustained ultimately not by engineering excellence, production numbers, or parts availability, though all those factors matter. It's sustained by people, clubs, events, online groups, and informal connections that continually renew interest, share knowledge, and welcome new generations into the community.

The enthusiast ecosystem operates at multiple scales and levels of formality. National organizations, such as the National Corvette Restorers Society and the Camaro/Firebird Nationals, provide structure, legitimacy, and large-scale events that draw participants from wide geographic areas. Regional and local clubs, Chevrolet clubs, muscle car associations, and street rod organizations create more accessible entry points with regular local meetings, cruise nights, and modest shows. Marque-specific groups focus on particular vehicle lines, accumulating specialized knowledge and resources relevant to those specific applications.

Events form the community's visible face. Major car shows display preservation efforts and inspire new projects. Cruise nights offer low-key social gatherings where cars are actually driven rather than just displayed. Track days and racing events showcase performance and technical development while fostering friendly competition. Swap meets facilitate the circulation of parts, connecting people with surplus components with those who need them. Engine-building workshops and technical seminars transfer hands-on knowledge in structured educational settings.

Digital communities increasingly complement and sometimes replace geographic organizations. Online forums create persistent repositories of technical discussions, build discussion threads, and provide troubleshooting advice accessible from any location. Facebook groups and similar social media spaces provide real-time interaction and quick responses to questions. YouTube channels document builds, demonstrate procedures, and preserve visual knowledge that text struggles to convey. Podcasts offer longer-form discussions of topics, culture, and technical details while people commute or work in their shops.

Why does community matter beyond just social enjoyment? Several practical benefits emerge from active participation. Communities dramatically reduce the learning curve for newcomers. Instead of reinventing solutions to common problems, new enthusiasts access collective knowledge representing millions of hours of combined experience. Someone's already figured out which brake master cylinder works best in that particular swap, which fuel tank sender matches modern gauges, or what causes that mysterious vibration at certain RPM ranges.

Communities create shared troubleshooting resources. When you're stuck on a problem, access to experienced builders who've encountered similar issues proves invaluable. Describe symptoms in

The History of the Small Block Chevy Motor

a technical forum, and responses often come within hours from people thousands of miles away with direct relevant experience. That distributed expertise network provides support impossible for any individual to match.

Communities help keep parts, tools, and knowledge circulating rather than disappearing. The retired builder selling decades' worth of specialized tools finds buyers who'll actually use them. Rare parts get traded between people who need different items. Technical knowledge is transferred through teaching, documentation, and collaborative problem-solving rather than dying with individuals.

Importantly, the Small Block community now spans age groups, income levels, backgrounds, and cultures in ways that transcend stereotypes. The expensive trailer-queen restoration exists alongside the budget-built street machine. Professional engine builders interact with first-time home rebuilders. Young enthusiasts bringing fresh perspectives work alongside veterans with decades of experience. International participants, from Australia to Brazil to Europe, contribute viewpoints and approaches shaped by different automotive contexts. Everyone, regardless of experience level or project scope, has something to contribute.

Consider a specific example: a regional Chevrolet club hosting an annual "Small Block Day." The event includes dyno sessions where participants can test their engines and compare results. Technical seminars cover topics from basic rebuilding through advanced tuning. A show-and-shine display showcases diverse Small Block applications, from original restorations to wild customs. Vendors provide access to parts and suppliers. But beyond the formal programming, informal conversations in the parking lot, shop tours, and connections made between participants often prove most valuable. Someone discovers a machinist who understands their rare engine variant. A first-time builder finds a mentor willing to guide them through their project. Parts change hands between people with complementary needs.

Another case study: an online forum thread documenting a novice's first complete Small Block rebuild. The builder posts regular updates, progress photos, questions when confused, and problems encountered. Experienced forum members chime in with advice, warnings about potential mistakes, and encouragement. Over months, the project progresses from a disassembled pile of parts to a running engine. The thread becomes a permanent resource, documenting not just success but the realistic challenges and problem-solving process. Other beginners reading the thread later gain confidence seeing someone else navigate their first build successfully.

Conclusion

The Small Block Chevy's story, from Ed Cole's postwar vision through today's crate LS builds, illustrates how thoughtful engineering can outlast any single era of technology. This engine did more than power cars; it embodied a design philosophy that proved durable across generations of vehicles, regulatory changes, and cultural shifts. That enduring legacy emerges from multiple interconnected factors we've explored throughout this chapter.

The production legacy speaks to unprecedented scale and sustained trust. When tens of millions of engines share a core architecture across seventy years of continuous production and evolution, the result is infrastructure, parts, knowledge, and a community that transcends any single application or era. That installed base creates its own momentum, making Small Block projects more accessible, less risky, and better supported than alternatives with slimmer production numbers or shorter lifespans.

The LS and Gen V continuation demonstrate that the fundamental philosophy, compact packaging, pushrod simplicity, robust bottom end, and modular adaptability remain viable in a modern, electronics-rich, efficiency-focused context. These aren't entirely new engines that happen to share a name; they're genuine successors that carry forward core principles while incorporating everything learned over decades of development. The bloodline continues, adapted but recognizable.

Crate engines and factory support transformed the Small Block from a vehicle component into a modular performance product. The engine's lifecycle no longer depends on OEM installation in specific vehicle models. As long as enthusiast demand exists, factory-backed crate engines remain available, warrantied, documented, and tested, enabling projects that range from straightforward restorations to cutting-edge restomods. That ongoing support legitimizes performance culture and ensures that future generations can access proven technology rather than scrounging increasingly scarce used examples.

The engineering lessons embedded in the Small Block's design, modularity enabling incremental improvement, simplicity reducing points of failure, adaptability accommodating diverse applications, and manufacturability ensuring accessibility, remain relevant regardless of specific technology. Whether designers are creating next-generation internal combustion engines, electric powertrains, or completely different systems, the principles that made the Small Block successful still apply. Sound fundamentals, thoughtful modularity, and realistic margins create technology that can evolve rather than become obsolete.

Environmental adaptation demonstrates that emissions compliance and performance aren't necessarily opposing goals. The Small Block evolved from pre-emissions simplicity, through crude smog equipment, toward modern direct injection and variable valve timing, continuously adapting to stricter regulations while improving both power output and efficiency. That progression required genuine engineering effort and added complexity, certainly. But it proved possible without abandoning the fundamental architecture or sacrificing the engine's essential character.

The electric future creates a context shift rather than immediate extinction. V8s are transitioning from mainstream powerplants to enthusiast-focused technology. That's change, absolutely, but it's not disappearance. The millions of existing Small Blocks, the supporting parts ecosystem, and the community of people who value this technology won't vanish simply because new vehicle production shifts toward other powertrains. The Small Block's role is evolving, becoming more specialized, more enthusiast-centered, but that evolution can sustain the platform for decades to come.

ABOUT THE AUTHOR

Todd Bandel is an accomplished author specializing in informational history books, particularly with a focus on the automotive industry. Drawing from 40 years of experience as an automotive technician, Todd combines deep expertise and passion to enlighten readers about the historical nuances of automobiles. Todd currently resides in San Diego, California, where he continues to explore and write about his enduring interest in automotive history.

Mechanicaddicts.com

www.ingramcontent.com/pod-product-compliance
Lightning Source LLC
Chambersburg PA
CBHW020651220526
45464CB00001B/389